ROBOTICS RESEARCH AND TECHNOLOGY

RECENT ADVANCES IN ROBOT PATH PLANNING ALGORITHMS

A REVIEW OF THEORY AND EXPERIMENT

ROBOTICS RESEARCH AND TECHNOLOGY

Additional books and e-books in this series can be found on Nova's website under the Series tab.

ROBOTICS RESEARCH AND TECHNOLOGY

RECENT ADVANCES IN ROBOT PATH PLANNING ALGORITHMS

A REVIEW OF THEORY AND EXPERIMENT

HADI JAHANSHAHI
NAEIMEH NAJAFIZADEH SARI
VIET-THANH PHAM
ROYA KHAJEPOUR
AND
CHRISTOS K. VOLOS

nova
science publishers
New York

Library of Congress Cataloging-in-Publication Data

ISBN: 978-1-53616-795-5

Published by Nova Science Publishers, Inc. † New York

CONTENTS

Preface

The scope of robotic science and its real-world applications have steadily flourished due to the ever-increasing demand for safety and efficiency. Generally speaking, the survey on the different types of robots can be mainly undertaken in three classified categories, namely design and manufacture, navigation and guidance, and the applications of robots. To put another way, the robot should be designed and evolved as a first thing; after that stage, navigated and guided, lastly operated for multifarious purposes on different occasions. The robot's mission is practically irrespective of the robot's role in its navigation and guidance. Due to the extensive breadth of research in the field of robotic navigation and guidance, it can be divided into four main branches, including positioning, path planning (routing), motion planning (control), and mapping.

In order for each robot to fully succeed in a mission, it requires a series of so-called instructions and rules as robot path planning and navigation. Routing has been of prominent importance throughout the history of robotics. Preliminary to the emergence of robots, two prolonged controversial issues, the "Travelling Salesman Problem" and "Piano Mover's Problem," have been debated and addressed in manifold strategies. Routing is broken down into three distinct categories corresponding to the provided information from different environments, namely routing in a known, semi-known, and unknown environment.

This book embraces the central concept of introducing the separate routing methods and putting forward some of the most proper ones for robotic routing schemes. These methods would create an ideal opportunity to be adopted in a vast variety of robots and to be resistant to disturbances. The further features that will be pointed out in the introduction of the methods are being real-time, autonomous, risk management, and having the ability to identify high-risk areas.

The introduction of the profound significance of the robots and delineation of the navigation and routing theme is provided in the first chapter of the book. The second chapter is concerned with the subject of routing in unknown environments. In the first part of this chapter, the family of bug algorithms including Bug1, Bug2, ALG1 and ALG2, DistBug, Tangent Bug and few more are described. In the following, several conventional methods, including D^*, Com, Class, and Rev algorithms, are submitted. The last part of this chapter is dedicated to the introduction of two recently developed routing methods; "Histograms of Vector Fields" and "multi-strategy approach."

In Chapter 3, routing is reviewed in the known environment in which the robot either utilizes the created maps by extraneous sources (e.g., the GPS) or makes use of the sensor in order to prepare the maps from the local environment. There are three pivotal policies on leading the robot towards the target, namely feature-based guidance, guidance with potential field and histograms of vector fields. The third approach is particularized in the second chapter and, the two first strategies are introduced in Chapter 3.

The robot path planning relying on the robot vision sensors and applicable computing hardware are concentrated in the fourth chapter. The first part of this chapter deals with routing methods supported mapping capabilities. In this section, three mapping methods such as polygonal, gridding, and hierarchical maps are detailed, and therefore, their merits and downsides are inspected. The second part manages the routing dependent on vision sensor typically known as the best sensor within the routing subject. The movement of two-dimensional robots with two or three degrees of freedom is analyzed within the third part of this chapter. Two standard methods, configuration space as well as the potential field are briefly encapsulated and based upon them, a combination algorithm with fast speed and high reliability is introduced, and its execution results in various circumstances are critically evaluated.

In Chapter 5, the performance of a few of the foremost important routing methods initiating from the second to fourth chapters is conferred regarding the implementation in various environments. The first part of this chapter is engaged in the implementation of the algorithms Bug1, Bug2, and Distbug on the pioneering robot. In the second part, a theoretical technique is planned to boost the robot's performance in line with obstacle collision avoidance. This method, underlying the tangential escape, seeks to proceed the robot through various obstacles with curved corners (obstacles whose shapes are U, V, Z, or zigzag). In the third and fourth parts of this chapter, path planning in different environments is preceded in the absence and the presence of danger space. Accordingly, four approaches, named artificial fuzzy potential field, linguistic technique, Markov decision making processes, and fuzzy Markov decision making have been proposed in two following parts and enforced on the Nao humanoid robot. As a conclusion, the results of those methods are provided.

Chapter 1

Introduction and Overview

One of the long-standing contentious questions between the scholarly world and industry in the field of robotics among specialists and activists, which have been inexorably heated controversy, is consistent with the definition of a robot. Every so often, this definition may emanate from a technological philosophy or an industrial expletory knowledge contingent on the manufactured robots defined by a business corporation. As an example, Robotic Industrial Association, an illustrious company in the field of industrial robot arms, offered an experiential defection for a robot as follows:

> "A robot is a reprogrammable, multifunctional manipulator designed to move material, parts, tools or specialized devices through variable programmed motions for the performance of a variety of tasks."

As reported by the International Standard Organization (ISO), Document 8373 (ratified 2012), Robot is specified as follows:

> "An actuated mechanism programmable in two or more axes with a degree of autonomy manipulator within its environment, to perform intended tasks."

Besides, according to the document, autonomy is defined as follows:

> "The ability to perform intended tasks based on current status and sensing, without human intervention." In accord with this definition, a block diagram for a robot can be represented as Figure 1.

Hidden Mafia made in the 1950s and 1960s by George de Waal and Joseph Engelberger, was one of the first robots to be created. Engelberger established the first robot-making company under the name Roboband, so for this reason, the title "father of robotics" was bestowed on him. Robots can be categorized in numerous methods, the most common two of which are inspired by the body-style and performance of the robot. This classification is depicted in Table 1, according to which humanoid robots are taken into account as the most characteristic types. Nowadays, the robotics researchers take special heed of humanoid robots. This approval is on the grounds that not only did these robots appear to be capable of matching and adapting themselves with human's living conditions, but they also possess the ability to cross the obstacles in housing arrangements, due to their better maneuverability in comparison with the other types. Despite the fact that plenty of researches have been being conducted in this regard expecting to build artificial limbs for humans and suchlike, there remains a long-distance journey to put servant humanoid robots on the commercial market by virtue of their extraordinary complexities.

In Table 1, BEAM robots can be controlled through making the most of biological and behavioral characteristics of organisms. BioRobots are frequently modeled based on the structure of living species in nature and given the fact that the creation of an insect- form robots is mechanically straightforward; these robots often share a pronounced affinity with insects. Creation of these robots has increasingly become popular special thanks to their simple mechanical movements, in turn demanding low power and less complicated calculations. Automated robots require electronic circuits just the same as all other robots. Aesthetics robots have dazzling and classical appearance without the need for multi-stage printed circuits.

In the light of the fact that humans are taken into consideration as one of the most sophisticated creatures in nature, for humanoid robots are also the most complicated human-made ones. That is why every breakthrough achieved in this field has an enormous potential to be utilized in order to pave the way for other technologies among which artificial prostheses can be pointed out. Mahro, Atlas, Kobian-R, and HRP4 which are produced by Korean Institute of Science and Technology, Boston Dynamics Company, Waseda University and Kavada industries, respectively, are in the midst of renowned humanoid robots. These robots are pictured in Figure 2.

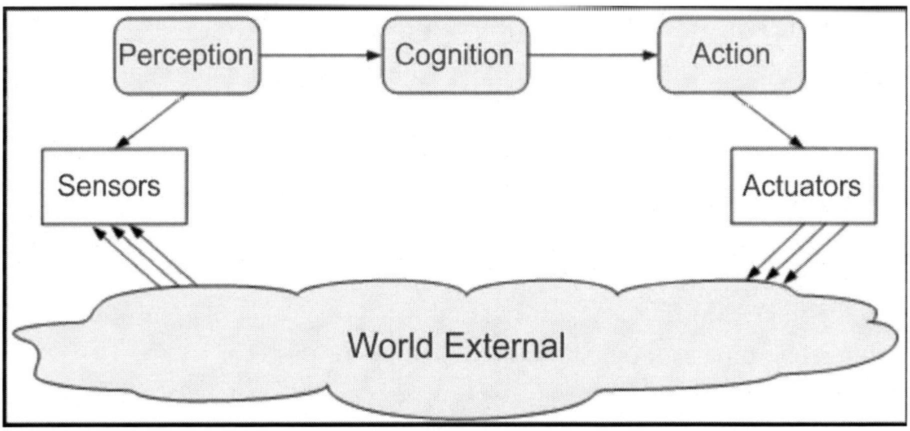

Figure 1. Block diagram representation of a robot in term of the 'autonomy' definition.

Aside from industrial robots, small humanoid robots are manufactured for laboratory objectives, which among the most overriding ones are; Nao, Zeno, and Darvin-OP (products of Aldebaran Robotics (France), Hanson Robokind and Robotis, respectively). These types of humanoid robots are displayed in Figure 3.

The momentousness of robots is supposed to be sought in their application as well as their playing a part in the present and future time. Specifically, humanoid robots will be the most used ones in the foreseeable future and will make a notable contribution to manufacturing, military, aerospace, medical, and other similar industries. Currently, the leading corporations all around the world are casting about for substituting humanoid robots for traditional manipulators. For example, a German company, DLR, laid the foundation for the first humanoid robot named Rollin Justin working on production sectors. Since 2013, America's military industries have put humanoid robots for ongoing wars into exploitation and operation stage. As shown in Figure 4, Atlas Robot which was ordained by Defense Advanced Research Projects Agency (DARPA), is known as the first robot that was devised and contrived by Boston Dynamics to be served in the military. Getting a grip on the evolution of impending wars affairs reveals that robots, particularly humanoid robots are going to revolutionize every facet of living condition in the near future higher than ever. Apart from military contexts, humanoid robots have breezed into the space industry. Viability of dispatching humans into space is open to debate. Therefore, National Aeronautics and Space Administration (NASA) has amended its grandiose plans and proclaimed that several humanoid robots would also cooperate with humans on their mission to the moon, henceforth, in order to downsize its workforce. By now, this administration has launched humanoid robot Robonaut II (Figure 5) to the international space station, and it is going to adopt the resolution to ship it to a thousand-day mission to the moon.

Table 1. Two common classifications of robots based on configuration and performance

Categorization based on configuration	Categorization based on Performance
flying robot	mobile robot
reptile robot	• rolling robot
robot fish	• walking robot
warrior robot	static robot
soccer robot	autonomous robot
humanoid robot	control robot
robot minesweeper	BEAM (bio, electronic, aesthetics, mechanic)
service robot	virtual robot

Atlas from Boston Dynamics

Mahru from Korean Institute of Science and Technology

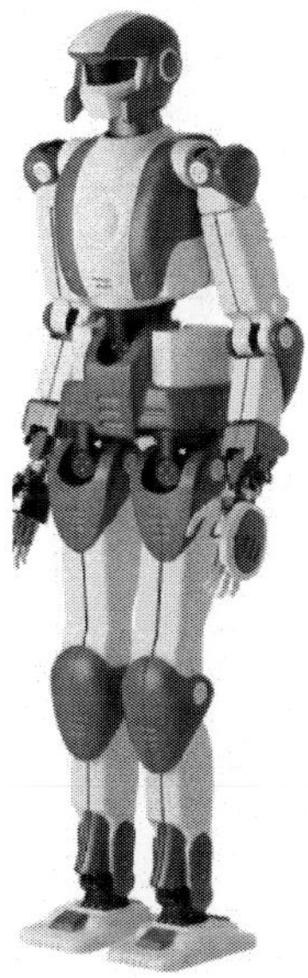

HRP-4 from Kavada industries

Kobian-R from Waseda University

Figure 2. Examples of human-size humanoid robots.

Zeno from Hanson Robokind

Nano from Aldebaran Robotics

Darvin-OP from Robotis

Figure 3. Examples of small-size humanoids.

Figure 4. Rollin' Justin from DLR.

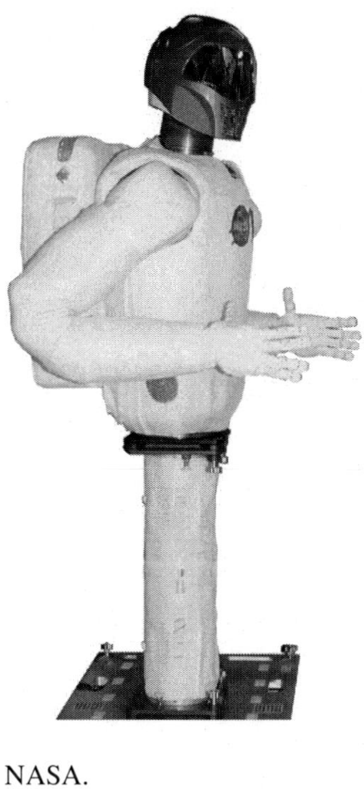

Figure 5. Robonaut II from NASA.

ROUTING, AN ONGOING CHALLENGE IN ROBOT GUIDANCE

On a large scale, researches generally in the field of robotics, in particular, humanoid robots, can be falls into three categories:

- Design and manufacture,
- Navigation and guidance,
- Application.

Otherwise stated, at first, the robot should be designed and constructed, and then, the navigation and guidance system tailored to their applications is supposed to be set up and ultimately these robots can be utilized in diverse environments for a wide range of applications and perform their role of navigation and guidance without being affected whether it is employmed at home or on agricultural fields. Simply put, applying these robots under any circumstances necessitates navigation and guidance. Owing to the significant proportion of robotic researches in the field of navigation and guidance, it can be divided into four main categories:

- Positioning,
- Path planning (Routing),
- Motion planning (Control)
- Mapping.

By and large, with the objective to conduct navigation and guidance, a robot should carry on figuring positions whose description can be either quantitative (to be specific, coordinates in terms of unit length) or qualitative (e.g., color patterns and so on). After this stage, the robot should determine a path dependent simultaneously on their current position, perception from the environment (with or without a map) and also the terminal position (target). Consequently, "path planning" is the task of assessment of the path through the environment that enables the robot to meet its preferable target. Only after the path identification, the robot is supposed to be controlled such that it is capable of going along this path. Robot motion control alludes to the motion planning methodology. Some sort of robots, primarily humanoid robots draw a map of their surrounding environment so that they would benefit from using this map for their ensuing routing targets. With reference to the robot's information provided from its environment, routing problems comes into three broad categories;

- Routing in a known environment,
- Routing in a pseudo-known environment,
- Routing in an unknown environment.

Regarding the case of the known environment, making good use of the full range of vision provided by a single camera or a set of cameras, a comprehensive map or a wealth of accurate information related to the route is given to the robot. Hardly had a complete map or broad vision been available for a given robot; it was practically impossible except for certain conditions in most cases. Extensive investigations have been undertaken in this realm, present-day routing dimensions in known environments are fully pinpointed and addressed.

In a pseudo-known environment, although the robot will be given a map of the environment as a first step, piece of information about a couple of dynamic obstacles (incl. humans or animals) or minor changes in the plan does not exist in the map at the mercy of lack of ability to save data. In spite of this situation, the pseudo-known environment also can serve analogous to the known environment, but when subjects to changes, the map will be updated.

In an unknown environment, the robot initially has no map and receives its information from sensors instead. In this case, concerning the type of robot hardware, routing can be conducted according to Table 2.

The capability to draw an accurate and detailed map of where the robot is assumed to proceed a short distance could be substantiated. For moderate distances, the tactic of making the map of the explored path sounds affordable. Be that as it may, over great distances owing to the necessity for colossal memory just as expanded processing workload for locating the optimal route, mapping is not virtually conceivable.

The strategies that are exclusively anchored in implementing the attendance sensors are obsolete, for these sensors lengthen the route and supply the robot with the limited data. Provided that distance sensors give reliable information for steering, but the data acquired from these sensors cannot offer robots with the knowledge of the danger zones. Additionally, the utilization of these sensors amid war (since being active and easily distinguished by the adversary) is not prudent. Visual sensors (cameras) are the leading sensors within the context of routing, given their provided information. All the humanoid robots need no less than one sensor for vision so as to obtain the actual distance, having double or multiple vision is still desired.

The structural motif of this book is to present various strategies of routing and to propose the ideal appropriate ones for robot navigation. The accompanying conditions prevailing in these methods are:

- Ability to run on an assortment of robots;
- Resistance to natural turbulence (noises);
- Being real-time;
- Self-regulation (autonomy);
- Ability to recognize high-risk areas and risk management.

Table 2. Robot routing methods in unknown environment

Classification method	Classification			
Based on sensors	visual sensor	distance sensors	attendance sensors	data fusion
	single vision	ultrasonic sensor		-
	dual vision	lasor sensor		
	multiple vision			
Based on drawing ability	without map (without memory)	mapping the travelled path	mapping the environment	

Chapter 2

PATH PLANNING IN UNKNOWN ENVIRONMENTS

INTRODUCTION

It has been a long time since routing entering into the realm of robotics science. Well ahead of the advent of robots, attention was oriented towards the set of two key challenges termed as "Street sellers "and" "Piano Displacer" over many years and their solutions have been suggested in a variety of methods. The routing was primarily examined in robotic manipulators and then in other subjects such as wheeled and flying robots. Since the emergence of humanoid robots, thus far, there has been proposed a certain number of approaches in order to route them, each with its strengths and weaknesses. The deep breath of routing applications is to such a degree that it is yet considered as an open issue.

In this chapter, two terms currently being used in the robot navigation should be mentioned at the outset:

Guidance: The use of this term, in robotics, is couched in the movement of a robot from one point to another position on the obstacle-free path. In the guidance process, the robot can either be equipped with a predefined path to follow or deliver itself to the target point premised upon a sequential identification of the surrounding environment. In order to reach the best decision during a robot's guidance towards the target, it becomes highly dependent on sensor data, updating the current position and direction in most cases.

Path planning: Path planning entails data accumulation and evaluation of some paths along which one or more conditions are fulfilled (e.g., bypassing obstacles, producing the shortest path length, imposing minimum rotation for the robot, and ensuring the greatest safety). In a case a fixed path is taken for guidance, this route should usually be pre-specified.

Robot guidance techniques towards a target are contingent on the pursuing of two key questions:

1. Are the starting point and final position of the robot pre-defined?
   If the final position (target) is being determined beforehand, the coordinates of which can be given to the robot. Otherwise, it is needed for the robot to set off on a quest to detect its target point. Each time at which the robot traverse along the path, its position can be identified either by the robot its own or by utilizing the external instruments like a GPS receiver.
2. Are the obstacles pre-specified?
   Either the obstacles can be determined by providing the robot with the obstacle's coordinates or the robot scans them with the aid of its sensors.

In order for the robot to successfully perform its missions, its path should be designed beforehand. The pre-planned motion of the robot can be implemented in various situations as the movement of the robot in the small distances, drawing and cleansing the environment (functions in which the environmental spaces should be completely considered by the robot), warfare and espionage purposes that is increasingly concerned for the national security.

Due to the lack or insufficiency of information in unknown environments, the robot needs to be capable of designing the route by its own and pursue this path towards the target. On the basis of the foregoing, this chapter deals with path planning approach for robots in unknown environments and is going to put forward some recent routing methods along with conventional ones. The second section covers Bug algorithms family, particularly with regard to Bug 1, Bug 2, ALG 1 & 2, Distbug, TangBug algorithms and will present a couple of other members of this family [1]. Likewise, different conventional methods, such as D*, Com, Class, and Rio algorithms, will be propounded.

The final part of this chapter provides a full description of two novel routing techniques named as "Vector Filed Histogram" [2] and "Multi Strategic" [3]. Vector Filed Histogram is a long-established approach for both unknown and known environments, and it functions on the basis of gridding the environment and drawing a histogram graph. By the same token, the multi-strategic method resolves the routing issue in unknown environments by simultaneous utilization of multiple strategies.

BUG ALGORITHMS FAMILY

Robot guidance in the two-dimensional environment is formed from a starting point and a target point. The robot seeks to locate an obstacle-free path as long as proceeding from the beginning point towards the final position (target). As depicted in Figure 6, three examples of two-dimensional environments are provided. According to which, the robot requires to get started from the box-shaped starting position without a path with uncertainty or difficulty until the achievement of the circle-shaped target point.

Lumelsky was a trailblazer in path planning with conducting early works in this field [4-18]. Prior to this debut, navigation of the robot in unknown environments followed a tortuous route that was performed utilizing algorithms such as Pledge [19] and Tarry [20]. It was unfortunate that the Pledge algorithm appeared incapable of taking a robot to a specified position. Aside from that, applying these two algorithms led to the robot's traversed path became extremely lengthy. By all means, some unprecedented creative approaches in which it is of necessity to access information from their environmental settings are available.

Bug algorithms are the primary path planning strategies through which delivering the robot to the target point is guaranteed. These algorithms are highly convenient for robots with real-time performances (prompt and immediate performance).

The guidance of the robot in several environments with a scarcity of information from a starting point S towards a target point T is the prime objective of bug algorithms. In the case, that no path could be identified for the attainment of the target point, the algorithm reports "unattainability of the target point" as a result.

Bug algorithms would be executed in all kinds of robots equipped with tactile or range-estimation sensors that are provided with a method for the local position determination (for instance a length-calculator or a GPS system). The robot is able to detect its route towards the target point automatically because of implementing this algorithm. Lumelsky has employed this algorithm on the robot manipulators, end-effector of which are required to reach a specific position [14, 21, 22]. The movement of the robot manipulators are virtually the same as the motions of the mobile robots in an unknown environment, and the sole distinction between them is that the manipulator arms have a rigid base.

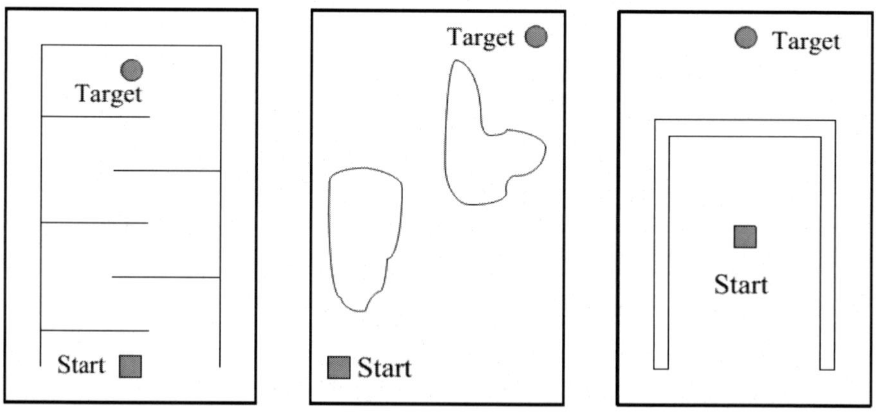

Figure 6. Some examples of 2D environments for robot guidance.

The further potential application of this algorithm can be scrutiny and questing an environment within a particular range to end in the target point [23]. Once the robot can locate the target's position, it approaches the target so that it is possible to inspect this point more precisely. One model for these robots is utilized in nuclear reactors in order to identify radioactive substances. If the robot detects a remote suspicious object, it is needed to draw near to it and take a full examination from the immediate vicinity. It entails specification of a strategy for the robot to proceed towards the suspected object in the environments consisting of a significant number of objects.

Sensors pioneered in Bug algorithms by Lumelsky and Skewis [12]. Utilization of sensors triggered the elaboration of the Bug 2 algorithm, which brought about shortening the length of the path through the creation of shortcuts. Fallowing after, there was raised a severe contention for the best point to bypass obstacles by robots; therefore, Kamon established "DistBug" Algorithm in order to accomplish this end. "TangentBug" is another algorithm with which the robot can employ some sensors to recognize the surrounding environments [24].

The key commonalities and discrepancies among the Bug algorithms reveal in the approaches that are adopted to ensure the accomplishment of their missions. SenseBug is a newly developed algorithm to be matched with sensors [25]. In this algorithm, three performance criteria are laid out viz frequency of data collection, path lengths, and the number of efforts to search. The strategy of this algorithm is based on the reduction in the frequency of gathering the available information in the environment and amounts of the search effort to aggregate data at any given time. Performing these special tricks will not hinder the mission from being achieved and accomplished.

Given the fact that the surrounding environment can constantly undergo changes, and there may be only a limited amount of information from it at each moment. Accordingly, the strategy for the robot's guidance should be formulated in a way that it can steer the robot towards the target point with a minimum quantity of information from the environment (ideally including both the target position and the robot's current position).

A few notable path design techniques, in particular, A* [26-28], Dijkstra [29], Distance Conversion [30-32], Potential Field [33-37], Sampling-based [38, 39], and piano stimulation problem [40-42] requires more data compared to two aforementioned strategies, and these approaches may occasionally necessitate a full map. These shortcomings draw attention that point-to-point guidance is inevitable in unknown environments.

What is more, path planning with bug algorithms has been utilized for robot guidance on the grassy area [43], crop harvesting [44], and for mining operations [45]. An algorithm titled as "WedgeBug" has been executed on mobile robots for surface exploration of the planet Mars by Laubakh and Burdick [46].

Langer, Coelho & Oliveira [47] came to conclusion that path planning in unknown environments could be implemented in a various major industries such as the health care (e.g., a surgical operation in a room which can be remotely controlled), construction, transportation, manufacture, military, aerospace (e.g., space research) entertainment (e.g., video games). In order to meet these burgeoning demands, they persisted in applying path planning for these different domains [41-45, 47, 48]. "K-Bug" Algorithm was simulated in an office room much like to real external environment and indicated that the planned paths via this algorithm could keep pace with the generated paths by the A* algorithm.

In short, Bug algorithms provides considerable advantages among which its simplicity and its capability to be perceived and then implemented stand out. At the minimum, these algorithms theoretically ensure the achievement of their target.

Indication and Terminology of Bug Algorithms

It is of critical importance to put forward the indications and terminology of bug algorithms family prior to the explications of these algorithms, some of the salient symbols are represented as:

S : Starting point

T : Target point, also stands for the end point.

x : Current position of the robot.

H_i : The ith collision point (among all the collision points) which indicates the ith time in all the number of times at which the robot strikes an obstacle boundary.

L_i : The ith separation point, which defines the ith time at which the robot separates from obstacle border and approaches the target.

$d(a, b)$: The geometric distance between arbitrary points of a and b.

$d_{path}(a, b)$: The length of robot's traveled path between two arbitrary points a and b.

r : Maximum range of the position sensors.

$r(\theta)$: The free space along the θ- direction between the robot and the first obstacle that intersects this direction.

F : Free space along the target. If the target is along with the θ-direction, F and r will be exactly the same thing.

In addition, the performances of all the aforementioned Bug algorithms for routing have been evaluated in the same theoretical environment to present the contrast between them (Figure 7). In this figure, the starting point and the target position are labeled with letters S and T, respectively. As will be shown anon, various algorithms with which to convey the robot from S to T, demonstrate markedly dissimilar and in some cases suboptimum routes.

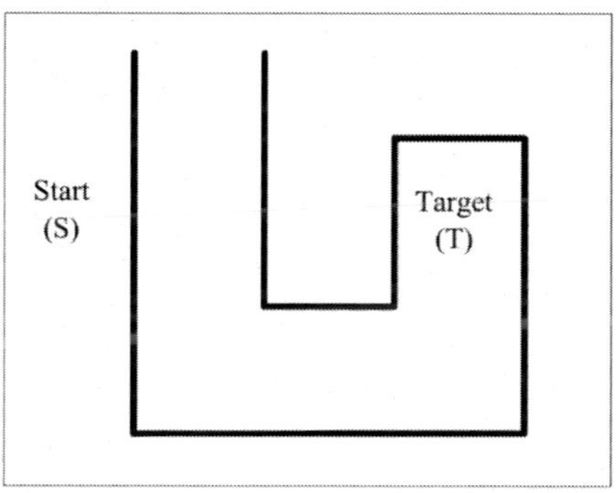

Figure 7. The theoretical environment to draw distinction between different Bug algorithms' performances.

Bug 1 Algorithm

Bug 1 algorithm founded by Lumelsky and Stepanov is the first algorithm to consider from Bug algorithms family [16-18]. According to the strategy of this algorithm, not until the robot encounters an obstacle, will it keep moving towards the target point. After coming across an obstacle, the robot circumvents it and selects the nearest location towards the target. In this algorithm, it is taken for granted that the target position is defined for the robot in advance, whereas the robot is not able to observe it directly.

The underlying principles behind the performance of Bug 1 algorithm are articulated as follows, and their demonstrations consistent with the mission are illustrated in Figure 8.

Step 0. Set $i = 0$ as an initial value.

Step 1. Increase i one unit and robot heads towards the target point until one of the outlined scenarios plays out:

Step 1.1. The target is achieved; the algorithm is successfully exited.

Step 1.2. An obstacle is encountered, the point at which the impact occurred is labeled "H_i" and then go to Step 2.

Step 2. Dodge the obstacle from left side in order to meet the point H_i, and comply with two conditions while searching for point (or points) L_i, viz., "the shortest distance to the target point" and "the likelihood to make a beeline for the target."

Step 3. Examine the traveling path to the target point

Step 3.1. If there exists a point L_i reach to point L_i and return to Step 1.

Step 3.2. If there exists no point L_i, the target is "unachievable," terminate the algorithm.

The robot in motion records the geometric length of the traversed path between the ith impact point H_i and its current position x as $d_{path}(x, H_i)$. Only points in which d has a minimum value and from which the robot can proceed towards the target are taken into consideration. Subsequently, the robot marks one of the minimum points (L_i), and as long as coming back to H_i and as long as coming back to H_i, it examines whether the target is attainable through moving towards the target from this point or not. If the target is inaccessible for the robot, the mission is aborted and failure will be reported as " the target is not attainable ". All the same, if the robot is able to achieve the target, the direction for navigating the obstacle boundary is chosen in such a way that $d_{path}(x, H_i)$ gets minimized. In consequence, the maneuver, at which L_i is attained, could be carried out and Step 1 is thoroughly performed.

The established pseud-code according to the strategy of this algorithm is encapsulated as follows:

```
While (TRUE)
  LET Len = line from S to T
  REPEAT
    Move from S towards T
    x = robot's current location
  UNTIL ((x == T) OR (obstacleIsEncountered))
    IF (x == T) THEN quit//target reached
    LET H = x//contact location
  REPEAT
  Follow obstacle boundary
  x = robot's current location
  LET L = intersection of x and Len
  UNTIL (((L is not null) AND (dist (L,T), dist (H,T) OR (x==T)
  OR (x==H))
  IF (x == T) THEN quit//target reached
  Move to L along obstacle boundary
  IF (obstacleIsEncountered at L in direction of T)
  THEN quit//target not reachable
ENDWHILE
```

Following Bug 1 algorithm, the maximum traversed path by the robot is quantified applying the following formula:

$$|\overline{ST}| = 1.5 \; [perimeter(obj_1) + perimeter(obj_2) + \cdots + perimeter(obj_n)] \tag{1}$$

in which S, T specify the starting and the end points respectively. A coefficient of 1.5 is worked out considering the robot bypasses the obstacle with a complete round for the sake of locating the best possible position to move off the obstacle border. After storing enough information, the robot is able to select the best place and return to this point in the remaining half round.

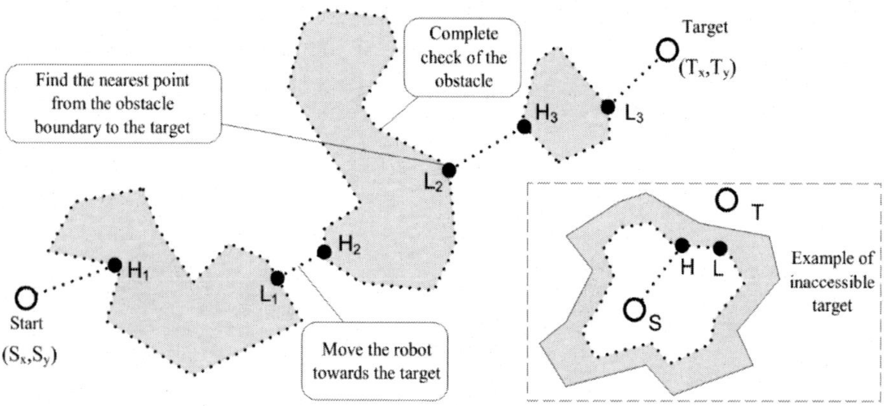

Figure 8. The general principle of Bug 1 routing method, for the two kinds of target; accessible and inaccessible.

In the Bug 1 algorithm, if the target is attainable, the achievement of the target point for the robot is guaranteed at all times. Besides, leaving the obstacle and moving towards the target is conducted at the best place by virtue of looking for the entire obstacle's border in a complete round. As a way of exemplification, the application of Bug 1 algorithm in a theoretical environment is depicted in Figure 9.

Bug 2 Algorithm

Bug 2 algorithm has also been set up by Lumelsky and Stepanov [16-18]. Not only is the algorithm Bug 2 inherently less conservative comparative to Bug 1 algorithm, but it does also leave the encountering obstacle's boundaries more readily. Whereas, in Bug 1 algorithm, the robot checks all the points of a barrier so that the nearest point to the target can be sought, performing the complete round in order to bypass the obstacle is hindered in the Bug 2 algorithm. If the robot is on the verge of hitting an obstacle, this strategy is draw up to navigate it to move along the direction of a straight line that is connected the start point to the target position, and in a case that it comes across any obstacle, it bypasses the obstacle from the left side.

In accordance with the hypothetical mission illustrated in Figure 10, principles for the performance of the Bug 2 algorithm are listed as follows:

Step 0. First, draw an imaginary line connecting the starting point to the target point and then initialize i with zero value.

Step 1. Add one unit to i, not until one of the following scenarios happens, do move along the line M towards the target point:

Step 1.1. The target is obtained, a successful termination of the algorithm.

Step 1.2. If an obstacle comes across, the collision point is marked as "H_i" and then run step 2.

Step 2. Get around the obstacle boundary on the left side until one of two possible scenarios is fulfilled as follows:

Step 2.1. If the target is successfully spotted, exit from the algorithm.

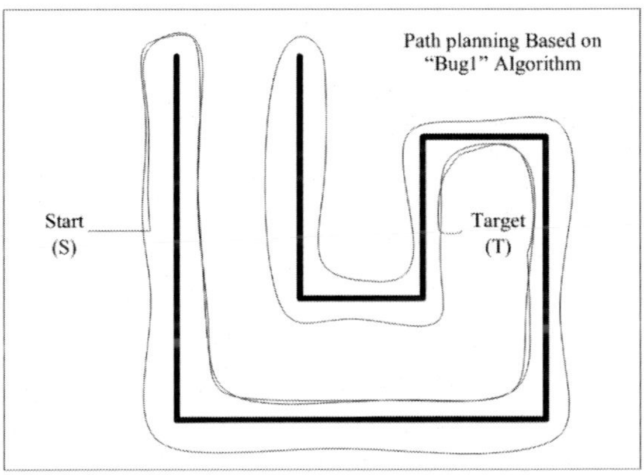

Figure 9. An example of the employment of Bug 1 algorithm in a hypothetical environment.

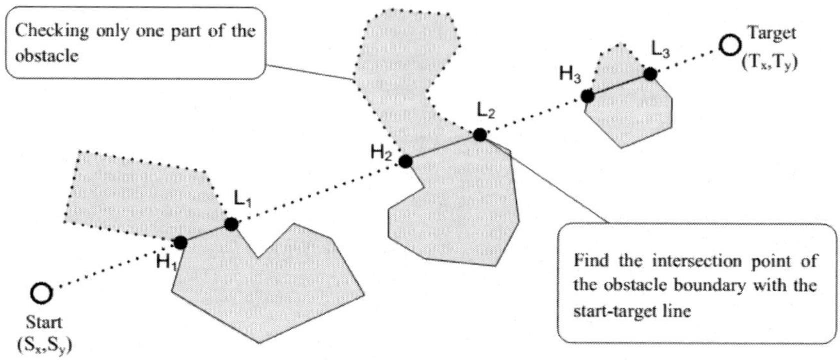

Figure 10. An example of Bug 2 algorithm's operation in a hypothetical environment.

Step 2.2. Some point along the line M should be found such that $d(x,t) < d(H_i,T)$. If movement towards the target is feasible, the point is labeled "L_i" and return to Step 1.

Step 2.3. If the point H_i is again encountered, stop the algorithm; the target point is unattainable.

The adopted pseudo-code for the evaluation of this algorithm is written as follows:

```
While (TRUE)
   LET Len = line from S to T
   REPEAT
      Move from S towards T
      x = robot's current location
   UNTIL ((x == T) OR (obstacleIsEncountered))
   IF (x == T) THEN quit//target reached
   LET H = x//contact location
   REPEAT
      Follow obstacle boundary
      x = robot's current location
      LET L = intersection of x and Len
   UNTIL (((L is not null) AND (dist (L,T), dist (H,T) OR (r==T) OR
   (x==H))
   IF (x == T) THEN quit//target reached
ENDWHILE
```

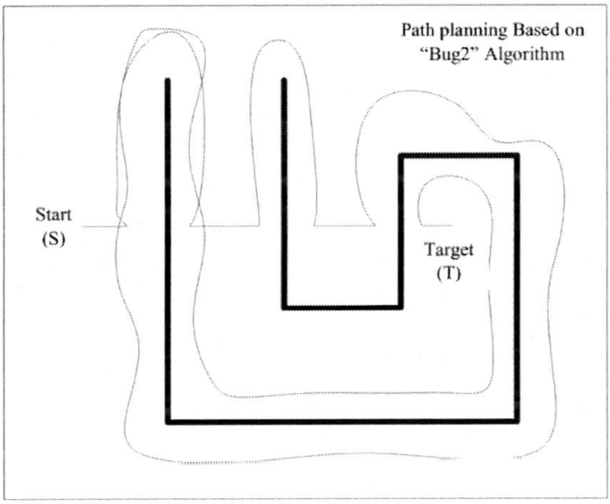

Figure 11. An example of Bug 2 algorithm's implementation in a hypothetical environment.

In a similar vein, if the target point is attainable, the target point with the robot's scanning can be definitely achieved in this algorithm analogous to Bug 1 algorithm. In order for the robot to identify the best place for leaving the obstacle's barrier, it embarks on an omnipresent and full-scale quest. The implementation of this algorithm in the desirable environmental settings is exemplified in Figure 11.

In Bug 2 algorithm, the maximal mileage for the robot (length of the traversed path) is computed via the following equation:

$$|\overline{ST}| = O\big(perimeter(obj_1)\big) + O\big(perimeter(obj_2)\big) + L + O\big((perimeter(obj_n)\big) \quad (2)$$

In view of the fact that the orientation at which the obstacle is bypassed (from the left or right side) is predetermined, and constant along the route, so the traversed distance of the robot in agreement with both the target and starting points might be either optimal or suboptimum. As evident in Figure 12, under the worst scenario policy (in which the robot takes a suboptimal path), the length of the traveled route is approximately on a par with that of the maximum possible distance. Another traversed path based on the suboptimal selection as an additional case is illustrated in Figure 13 during which the robot bypasses from the left to achieve the target, thus, it moves across a rather long route. However, had it bypassed from the right, it could have traversed a considerably shorter path.

It is obvious that Bug 2 algorithm performs more rapidly than Bug 1. It is worth mentioning that it is hypothesized that the robot in its close proximity is solely capable of recognizing the existence of only one obstacle in both of these algorithms. When the robot is equipped with obstacle-detection sensors in the 360 degrees angular range, it is overwhelmingly likely for Bug 1 and Bug 2 algorithms to be improved.

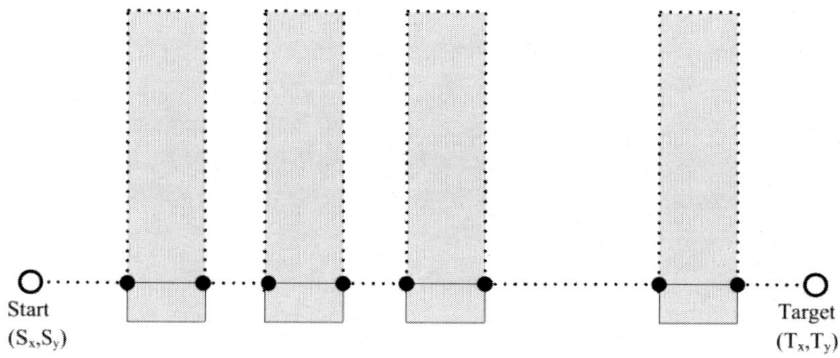

Figure 12. Traversing a roughly maximum distance due to the selecting suboptimal path by Bug 2 algorithm.

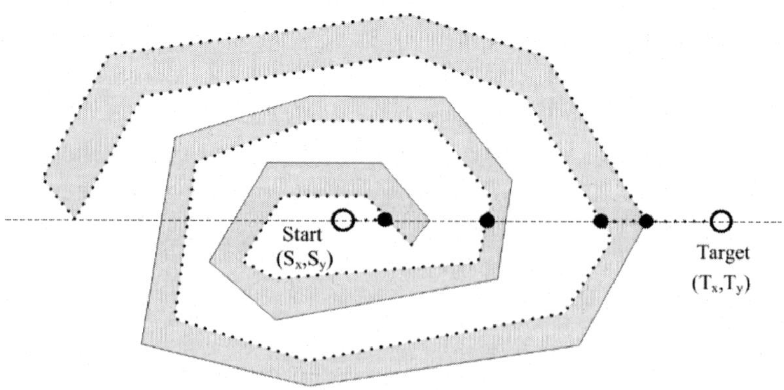

Figure 13. Turning to the left side to circumvent the obstacle (another suboptimal choice by Bug 2).

ALG1 Algorithm

The extended version of Bug 2 algorithm that was devised by Sankaranarayanan and Vidyasagar is widely known as ALG1 algorithm [49]. One potential shortcoming of the algorithm Bug 2 relates to the fact that likeliness of the robot's traveling path for one time or even several times is strengthened in this algorithm, resulting into the longer generated routes. With a focus on minimizing this weakness, ALG1 algorithm reminds the robot of obstacles' corresponding collision and leaving points to keeps them in its memory and utilizes them when needed for producing next shorter paths

The underpinning principle for the performance of ALG1 algorithms is outlined as follows:

Step 0. First, draw an imaginary line linking the starting point to the target position, assume an initial value of 0 for i.

Step 1. Add one unit to i, and the robot continues passing across the line M towards the target point until one of the subsequent scenarios takes place:

Step 1.1. The target is reached; satisfactorily quitting the algorithm.

Step 1.2. If an obstacle is encountered, the collision point is labeled "H_i", then go for running Step 2.

Step 2. Bypass the obstacle border on the left side in order that one of the following scenarios shows up:

Step 2.1. The target is located; the algorithm is successfully finished.

Step 2.2. One point as y with the correspondent features needs to be found as follows:

- The point belongs to the line M.
- For all the x's: $d(y,T) < d(x,T)$
- When the robot stands at point y, it is able to go towards the target point. Step 2.2.1. The resulting point is tagged as "L_i", and Step 1 is iterated.

Step 2.3. Travel in the proximity of point H_i and then come back to this point. If H_i is approached, walk along the obstacle's boundary and bypass the barrier wall from the left side.

Step 2.4. Return to the point H_i, in this case; the target is unattainable; the algorithm is over.

An illustration for the execution of this algorithm in a hypothetical environment is accounted for in Figure 14.

ALG2 Algorithm

The revised and developed version of ALG1 algorithm that was proposed by Sankaranarayanan and Vidyasagar is named ALG2 algorithm [49]. Contrary to the algorithms as mentioned earlier, the concept of line M (through which the starting point and the target point is connected) is no more defined in this algorithm, and different conditions are set up to overstep the obstacle by the robot. The operation principle of Algorithms ALG2 is explained as follows:

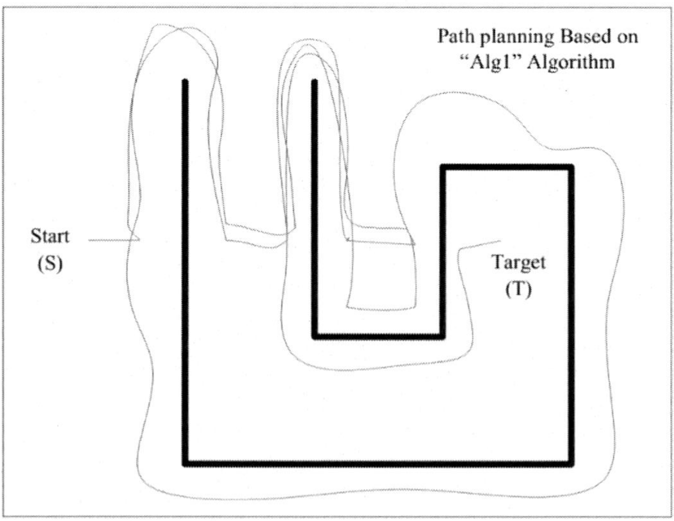

Figure 14. An example of the execution of ALG1 algorithm in a hypothetical environment.

Step 0. Two initial settings are $i = 0$ and $Q = D(S,T)$; Q refers to the nearest point of the robot to the target.

Step 1. Add one unit to i, proceed towards the target point, and update the amount of Q with setting the value of $d(x,T)$ simultaneously. Keep updating Q as long as the condition $Q < d(x,T)$ maintains. Further this process until one of the two underlying scenarios takes places as follows:

Step 1.1. If the target point is obtained, quit the algorithm.

Step 1.2. If an obstacle is detected, the point should be labeled as " H_i", and then Step 2 is run.

Step 2. Constantly hold the obstacle on the right side (meaning that bypassing the obstacle is started from left side) and simultaneous with this, continuously update Q with $d(x,T)$ until the condition $Q < d(x,T)$ maintains, this process keeps running until one of the following scenarios is true:

Step 2.1. The target is met; successful exit from the algorithm.

Step 2.2. The point y is met with the two constraints as follow:

- $d(y,T) < d(Q,T)$
- The robot is able to proceed towards the target from this point.

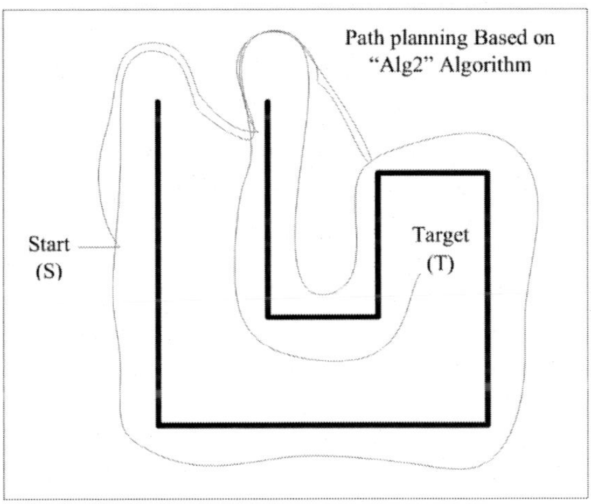

Figure 15. ALG2 algorithm's performance in an arbitrary environment.

Step 2.2.1. The found point is labelled" L_i", and return to Step 1.

Step 2.3. Turn around in the proximity of point H_i and then come back to it. If H_i is reached, bypass the obstacle from the left side.

Step 2.4. Return to point H_i; the target is unattainable and the algorithm is stopped.

As a case in point, the performance of this algorithm in an arbitrary environment is exhibited in Figure 15. It should be emphasized that the ALG2 algorithm has demonstrated an enhanced efficiency in comparison with the previously mentioned algorithms given the algorithms' criteria to leave a barrier thanks to not needed line *M* in order to fulfill this condition.

DistBug Algorithm

This algorithm was established by Kamon and Rivlin [50]. This algorithm applies a range-finder sensor for not only detecting free space *F* but also defining the conditions of leaving an obstacle. Performance of DistBug algorithm must meet the following requirements:

Step 0. Initialize i with the value of 0 and traverse a distance equivalent to the wall thickness (the minimum thickness of a barrier in the environment is considered for this displacement, and its amount must be given to the application by the user; this issue regarded as a weakness of DistBug algorithm).

Step 1. Add one unit to i and advance towards the target point to satisfy one of two conditions as below:

Step 1.1. The target is found, successful exit from the algorithm.

Step 1.2. An obstacle is encountered; this point is labeled "H_i" and Step 2 is run.

Step 2. Circumvent the obstacle from the left side and update the minimum amount of $d(x,T)$ at once and records it as $d_{min}(t)$. This process proceeds until one of the following conditions is accomplished:

Step 2.1. The target is observable, which means that the condition $d(x,T) - F$ remains. This point is labelled "L_i" and Step 1 is run.

Step 2.2. If $d(x,T) - F \leq d_{min}(T) - step$, this point should be labeled "L_i" and Step 1 is run.

Step 2.3. Take a loop and return to H_i; means that target is unattainable and the algorithm is stopped.

An example to show how this algorithm performs in a hypothetical environment is presented in Figure 16.

The TangBug Algorithm

The fundamental theorem of TangBug algorithm has been designed by Kamon, Rimon, and Rivlin [24]. The foundation of this algorithm is built on the incorporating distance-finding sensors in the map generating process from the local environment for the robot, and TangBug algorithm strives to shorten the route towards the target point as much as possible. The Ideal case would be when the robot's sensors can determine the distance of surrounding obstacles with respect to all of their corners and edges.

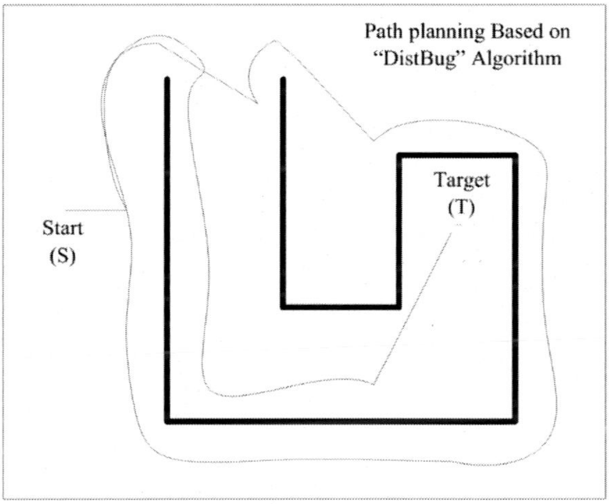

Figure 16. Illustration of the performance of the algorithm Distbug in an arbitrary environment.

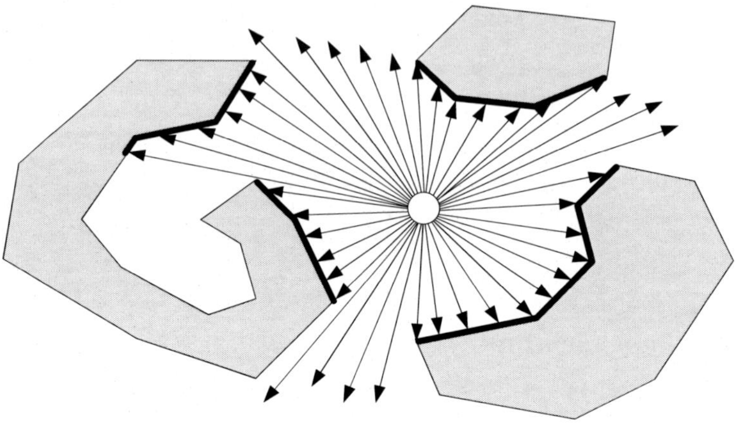

Figure 17. Obstacle recognition with the aid of sensors.

However, it is practically impossible for the robot's sensors to scan all the angles of obstacles so that the distance between the robot and barriers can be continuously and determined. discretely

To give an example, the distance between the robot and an obstacle is calculated after each rotation by 5 degrees. However, there could be several obstacles between these angle intervals somehow are outside the robot's vision applying this strategy. Given that the robot's sensors can only cover a specified range, let alone recognizing barriers beyond this range, leading to some obstacles cannot be detected.

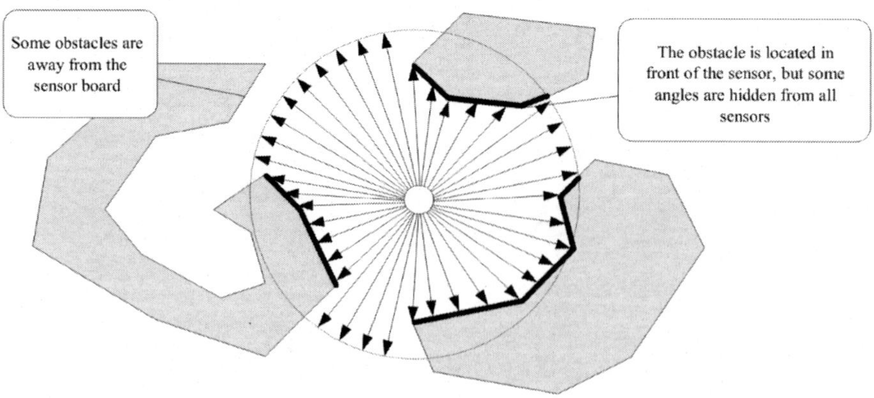

Figure 18. Obstacle recognition with the help of limited-range sensors.

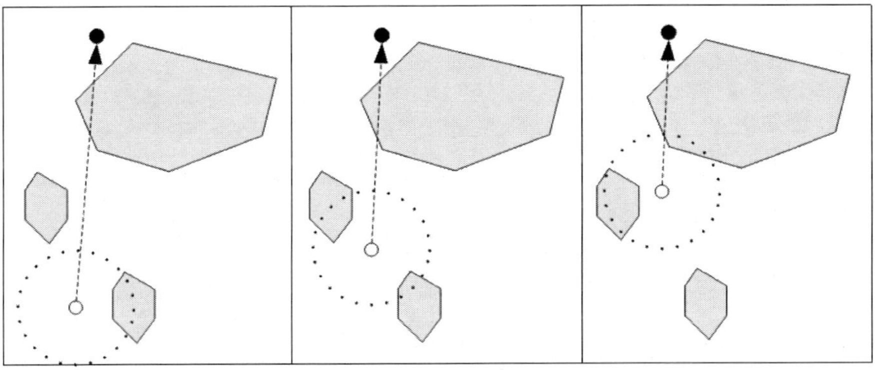

Figure 19. Robot's straight movement towards the target before sensing the obstacle.

The working principles of TangBug are formulated akin to Bug 2 algorithm:

Step 0. Approach the target point until sensors detect an object interrupting the direct path towards the target.

Step 1. Proceed towards the discontinuity points e_1 or either e_2 (Figure 20). The principle, to which the robot adheres in order to choose between these two points, is the selection of the shortest path to the target. It is made apparent in Figure 21, the robot selects its way according to this algorithm in such a manner that the amount $min(|\overline{Sc_1}| + |\overline{e_1T}|, |\overline{Se_2}| + |\overline{e_2T}|$ is procured.

Surround Tangential Chart

To begin with, an environment similar to Figure 22a that is composed of some obstacles with different shapes, a starting point, and a target is taken account. In the next stage, according to Figure 22b, convex vertices of each obstacle should firstly be determined. The specified vertices, the starting point, and the target position would be connected through straight lines based on the condition that these lines do not intersect the barriers. The resulting map demonstrating in Figure 22c is known as the surround tangential graph. It could be substantiated that this graph has the capacity to comprise the optimal route from the starting point to the position of the target. This optimal path for the under-consideration environment is depicted in Figure 22d.

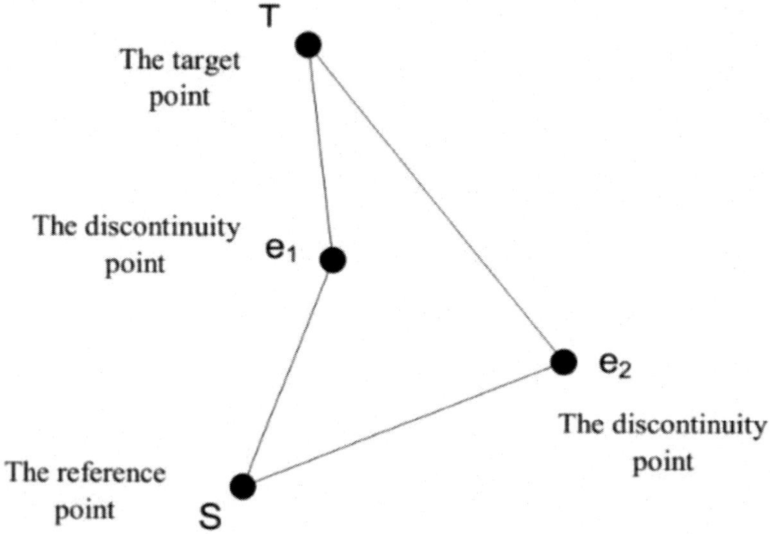

Figure 20. The start point, target point and two discontinuity points.

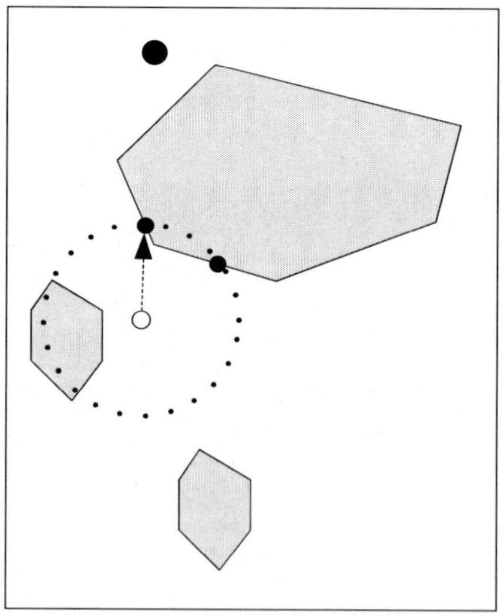

Figure 21. Robot's motion towards discontinuity points.

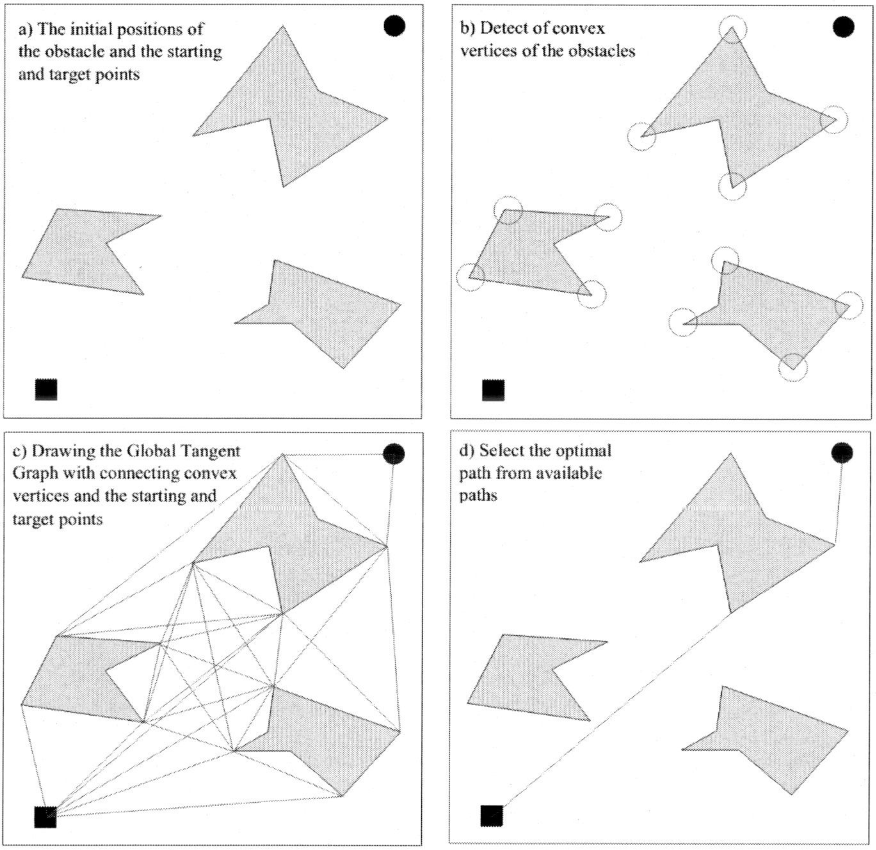

Figure 22. The stages of drawing the tangential diagram and selecting the optimal path in an environment comprising of obstacles and starting points and targets.

Local Tangent Graph

If the robot is incapable of extracting environmental information from its surrounding, it aims to draw a local map in return; this approach is referred to as the Local Tangent Graph (LTG). Figure 23 serves as an example for illustration of this graph. To generate the local tangent graph, it is required for the collected information to be applied in two functions $r(\theta)$ and F. The function $r(\theta)$ computes the distance to the first visible obstacle along direction of θ and by the same token the free space in front of the robot is quantified in function F. Utilizing this information, calculation of local tangent chart nodes and the optimal route is performed as follows:

A. Nodes resulted from examining status *F*:
   - If $d(x, T) \leq F$, the target is visible and a node named *T* is designated on it.
   - If $F \geq r$, this indicates that no obstacle is observed across the target point, accordingly, the node *T* is designated alongside the target. A case in which these nodes has been created, is demonstrated in Figure 23.

B. The nodes resulted from examining state of function $r(\theta)$:
   - By the time at which a discontinuity in the function $r(\theta)$ appears, a node along θ is made. An example of these nodes whose numbers are 1, 2, 3 and 4 is exhibited in Figure 23.
   - If $r(\theta) = r$ (implying that the distance of obstacle is equivalent to the maximal range of sensor) followed by the reduction of $r(\theta)$, a node will be added along θ. In Figure 23, node 5 is an instance of such nodes.
   - If $r(\theta) \neq r$, and thereafter, $r(\theta)$ becomes equal to r, a node is assigned along θ.

C Determining the optimal path:
   - For each node, the distance function $d(N_i, T)$ is determined, in which N_i and *T* stand for the ith node and the target node, respectively.
   - Any node that holds the lowest value of the distance function $d(N_i, T)$, is implemented as the optimal node and labeled "*N*".

Local Minimum

Figure 24 vividly demonstrate that the straight motion of robot towards the target leading to the shorter distance notwithstanding, the presence of an obstacle between the robot and the target makes a complete separation between them in certain circumstances. To put it differently, even though the robot is situated in a local minimum position, being stuck with an obstacle makes the robot away from the target once the robot's genuinely apparent approaching towards the target. This is precisely why some broader aspects of the environment should be considered in order to cause the robot to proceed towards the target point.

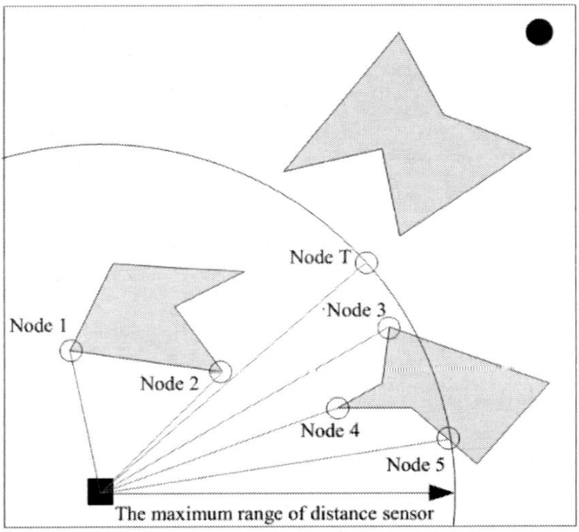

Figure 23. An example of localized tangent diagrams.

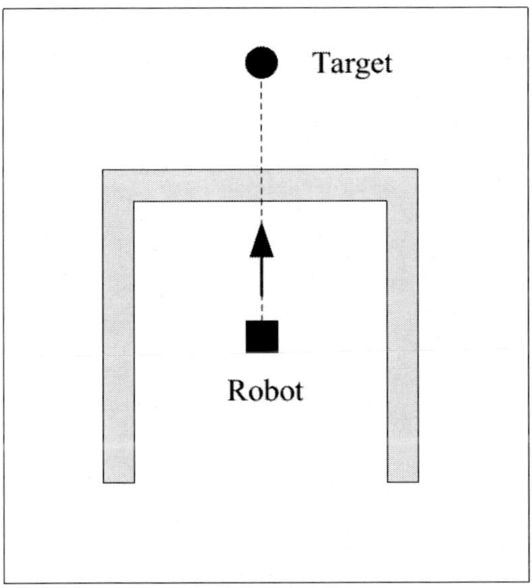

Figure 24. Robot getting away from the target due to local minimum position.

In a case when the robot is in a local minimum, TangentBug algorithm adopts the method of moving across the obstacle's boundary. In this situation, the robot picks one direction to circumvent the barrier due to the use of a graph of local tangential, and afterward, continuously updates two corresponding variables defined as below:

$d_{followed}(T)$: The minimum distance to the target along an obstacle with the lowest effect.

$d_{reach}(T)$: The minimum distance from point P to the target point.

At every moment, TangentBug algorithm monitors the available space for the robot and procure the point P at which the distance function $d(P,T)$ has its minimum value. Into the bargain, the amount of $d(P,T)$ supersedes the current value of $d_{reach}(T)$. The process of following the barrier's boundary prolongs until one of the following two conditions is accomplished:

- If $d_{reach}(T) < d_{followed}(T)$: the target is unattainable.
- If the robot entirely bypasses the obstacle and it has to perform this task again; similarly, the target is unattainable, and the mission aborts.

The pseudo-code for TangentBug algorithm in such an environment displayed in Figure 25 is expounded as follows:

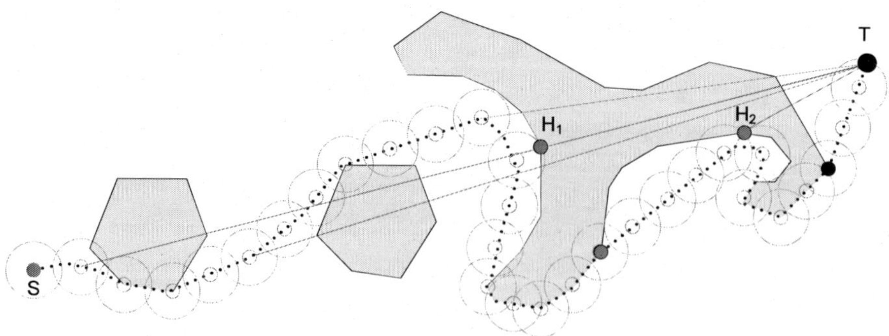

Figure 25. Routing strategy in TangentBug algorithm in an environment with multiple obstacles.

```
While (TRUE)
  LET w = T
  REPEAT
  x' = x//robot's previous location
    update x by moving towards w
    IF (no obstacle detected in direction w) THEN w = T
    ELSE
      LET eL and eR be discontinuity points
      IF ((dist (x,eL) + dist(eL,T)) < (dist (x, eR) +dist (eR,T)))
        THEN w = eL ELSE w = eR
   UNTIL ((x ==T)) OR (dist (x' , g) < dist (T,g)))
   IF (x ==g) THEN quit//target reached
   LET H = L = x//contact location
   LET dir = direction of continuity point (L or R)
   REPEAT
      LET w = the discontinuity point in direction dir
      IF (dist (x,T) , dist (L,T) ) THEN L = x
      Update x by moving towards w
   UNTIL ((dist (x,T), dist (L,T)) OR (x == T) OR (x == H))
   IF (x == T) THEN quit//target reached
   IF (x == H) THEN quit//target not reachable
ENDWHILE
```

The resultant trace of implementing TangentBug algorithm in an arbitrary environment is sketched in Figure 26.

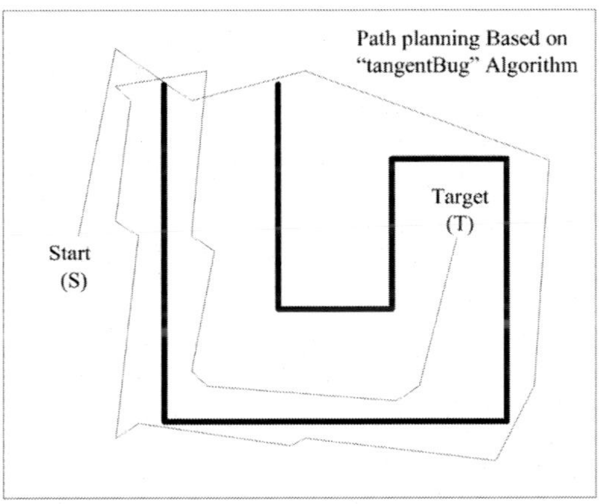

Figure 26. Implementation of the algorithm TangentBug in an arbitrary environment.

Other Bug Algorithms

Exclusive of the preceding algorithms, there exist several alternative algorithms that function based on Bug algorithm, each of which strives to optimize a special section of it implementing the auxiliary apparatus (e.g., sensors, GPS, camera, map, etc.). Indeed, these algorithms have not been at the core of ubiquitous attention; on this ground, it would suffice to introduce them in this section briefly.

HD-I Algorithm

Algorithm "HD-I" is analogous to Rio1 algorithm already mentioned in section 2-2-4. These two algorithms differ in the orientation in which they follow the obstacle wall. The HD-I algorithm is founded on the essential characteristics, namely the distance from the robot to the target, orientation, and the before followed directions. This method contributes to a high degree to minimize the path.

Ave Algorithm

Algorithm "Ave" that is an enhanced version of HD-I algorithm to be exact, presented by Noborio et al. [51]. The pivotal refined feature of this algorithm is to specify the direction along which robots move across the barriers' walls. In this algorithm, procurement of the distance to the target is based on not only current nodes' information but also the information of previously passed nodes by the robot.

VisBug 21 and 22 Algorithms

Algorithm VisBug 21, presented By Lumelsky and Skewis [12], enables the robot to follow the path established in Bug 2 Algorithm implementing sensors that have a definite range limit. With the aid of these sensors, finding shortcuts, which in return results in shortening the path, has been devised. VisBug 22 Algorithm has more risk-taking behavior to identify the shortcut in contradistinction to Bug 21 algorithm. This algorithm is efficient to decrease the length of path traversed by the robot dramatically; however, it is conceivable that this route would be lengthened if the shortcuts were not functional anymore.

WedgeBug Algorithm

WedgeBug Algorithm was put forward by Laubach and Burdick [46]. Wedges are analyzed in this algorithm, and the first wedge indicates the direction that conveys the robot to the desired target point. If no encounter between an obstacle and the robot occurs, it maintains its path towards the target, but if the case in which there exist any obstacle arises, the robot proceeds to follow the hypothetical barrier's border. Meanwhile, the vast majority of wedges are evaluated by means of limited-range sensors of the robot. The strength point of this algorithm is to obviate the need for generating either a local or global tangent graph. It should be born in mind that the creation of regionsl or surround tangent graphs are required in algorithms, for instance, TangBug algorithms. This algorithm is designed under the following assumptions that should always be checked:

- No information is available in advance;
- The plan should be performed based on the collected data from both sensors and cameras;
- The design should be robust, meaning it should predict a proper mechanism with respect to changes of the environment;
- The design should be comprehensive and accurate.

CautiousBug Algorithm

The logic of CautiousBug algorithm was proposed by Magid and Rivlin [52]. This algorithm comprises spiral motions, which leads the robot to successively switch the pursuit direction towards the obstacle.

Three-Dimensional Bug (3DBug) Algorithm

3DBug algorithm was offered by Kamon et al. [53]. Indeed, this algorithm is the refined version of TangBug algorithm that can operate in three dimensions rather than two dimensions. Multiple points of concerns, which are not handled easily compared with the robot's following the border of the obstacle, arises in this context since the robot scans in a large-scale environment.

Axis-Angle Algorithm

Lumelsky and Tiwari framed the general concept of Axis-angle algorithm [13]. This algorithm was motivated by the central principles of Pledge algorithm [19] and formulates two variables to get off the barrier, first of which is the angle that the connected line from point x (the current position of the robot) to the target point makes with the line from the starting point to the end point. The second variable is defined as the angle between the line connecting the beginning point to the target point and the current velocity vector of the robot. One of the outstanding advantages of this algorithm in contrast to the rest of Bug algorithms is that range measurement tools (including sensors) are not involved, for this reason; a merely simple direction indicator can be implemented to execute this algorithm.

Opitm-Bug Algorithm

Foundation of Optim-Bug Algorithm was made by Kreichbaum [54]. Contrary to other algorithms of Bug family, this algorithm employs an infrared sensor for the acquisition of environmental information, and it can simultaneously use this information in order to draw the map of the surrounding environment. When Optim-Bug algorithm has complete knowledge of the environmental situation, it could disregard the step in which the robot follows the barrier's border. When the robot is equipped with this sensor in addition to a map, this algorithm would be capable of selecting the shortest path, leading to a sharp decline in the amount of time the robot takes to reach the target without encountering obstacles.

Unknown Bug Algorithm

Unknown Bug algorithm [54] seems to be the same as Optim-Bug algorithm, but with one exception, that some sources of uncertainty are considered in the robot path. This algorithm quantifies the optimum route at each phase of the robot's motion and in order to do so, exploits available information (received from GPS, camera, maps, etc.) as well. Another motivation of this algorithm is to minimize the maximum error of the final position of the robot with regard to the target point.

SensBug Algorithm

The motive of developing SenseBug algorithm [25] is inspired by the need to corporate robots in building spaces such that the obstacles are supposed as simple objects with a fixed length. Leaving obstacles in this algorithm is comparable with that of COM algorithm, illuminated in the next section. It is acknowledged that obstacles can change their position in building environments. Based on this premise, it is confirmed that SenseBug can succeed in the robot guidance towards the target. Using the wireless network in this algorithm is imperative to navigate the robot in a given direction so that the maximal path length can be reduced during pursuing the barriers' borders.

D* Algorithm

D* Algorithms was proposed by Stentz [55], and it's explicit policies differentiate it from Bug algorithms, the most important of which is the employment of maps (mapping is not permitted in Bug algorithms). Consequently, D* algorithms demonstrate well-formed and unique features. The motion direction in this algorithm is specified through map gridding, and based on scoring these grids, as illustrated in Figure 28.

D Algorithm Performance without Barriers*

In order to gain a better understanding of the D^* algorithm, suppose a field with 5×5 grid considering Figure 28. In the initial state, the robot places in the cell (3, 1), and the target is set in the cell (3, 5). It is assumed that the step sizes for moving along horizontal and vertical direction are the same and weighted with number 1, and diagonal stepping with number $\sqrt{2}$. The algorithm D^* under this assumption makes Table 3 for the adjacent cells to the target cell (T). Firstly, in this table, determining the value of each cell near the target is made based on the stepping direction (including horizontal, vertical or diagonal) towards the target cell, afterward specifying the selected cells whose total value are lowest is performed. Considering both the Table 3 and Figure 28a in which the cells (4, 5), (2, 5) and (3, 4) possess the lowest total value, determining the direction towards the target.

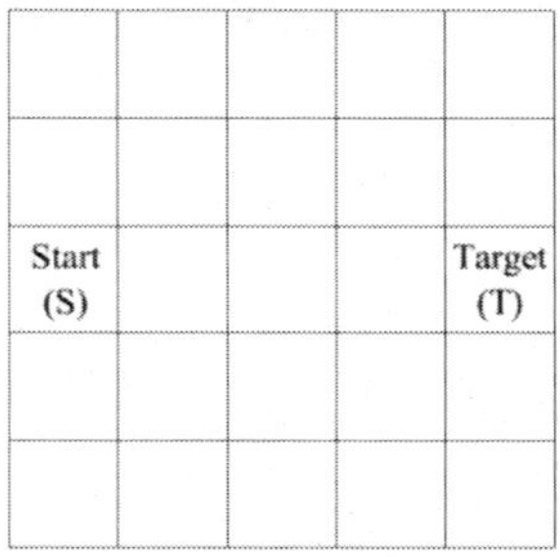

Figure 27. A simple example of gridding the environment using D* method.

In the subsequent stage, the cells in the neighborhood of the chosen cells (4, 5), (2, 5) and (3, 4) are valued on a similar pattern. The results of this valuation (i.e., value setting) are reported in Table 3 and Figure 28b. As evidenced, two cells (4, 4) and (2, 4) possess the lowest total value, and are determined as the selected cells to proceed the next phase and their direction towards the target is set.

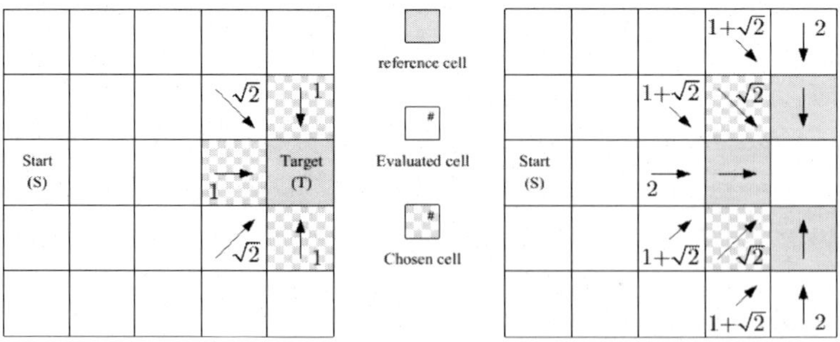

a) Set the direction of the target cell's neighbors

b) Set the direction of the neighboring cells of the chosen cells shown in a

Figure 28. Adjusting the order of the cells with the lowest value to the target cell according to the Table 3.

Table 3. D* algorithm for surrounding cells of target cell

Valued cell	Neighbor cell with at least distance to target*	Step value from (1) to (2)	Step value from (2) to (T)	Total value of cell
direction determination of neighbor cells of the target cell (Figure 28a)				
(5,4)	T	1	0	1
(5,2)	T	1	0	1
(4,3)	T	1	0	1
(4,2)	T	1.414	0	1.414
(4,4)	T	1.414	0	1.414
direction determination of neighbors' step 1 selection cells (Figure 28b)				
(5,5)	(5,4)	1	1	2
(4,5)	(5,4)	1.414	1	2.414
(4,4)	T	1.414	0	1.414
(3,4)	(4,3)	1.414	1	2.414
(3,3)	(4,3)	1	1	2
(3,2)	(4,3)	1.414	1	2.414
(4,2)	T	1.414	0	1.414
(4,1)	(4,2)	1.414	1	2.414
(5,1)	(5,2)	1	1	2

* The nearest cell to the target can be the target itself

This process continues until the host cell of the robot contains a pointer as well. The finalized process for the gridding example is displayed in Figure 29. Pointer-containing cells indicate the minimum value direction towards the target, tracing these cells results in a minimum value path towards the target.

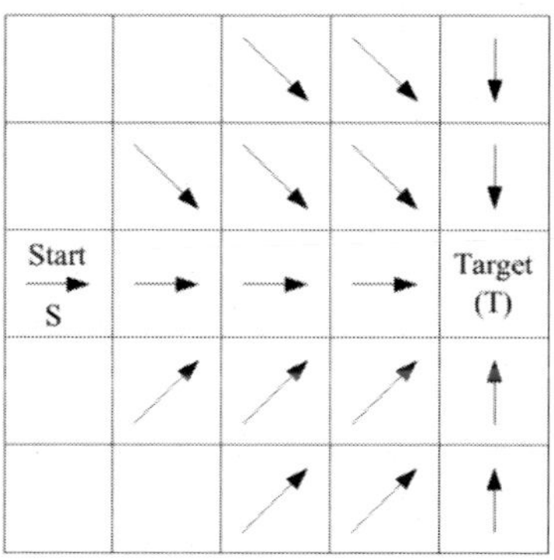

Figure 29. Completing the process of directing the field cells to get to the start cell (robot cell).

Case of Taking Obstacles in D^* Algorithm into Account

The strategy of the algorithm D^* for avoiding obstacles is designed to increase the value of movement towards the cells within which obstacles appear. The value-added contribution is made if only the obstacle-containing cells are approached and cannot be seen while leaving the cell. To illustrate, if there exists an obstacle in the cell (3,3) in Figure 30a, the value of stepping from the barrier's upstream neighbors (adjacent to the starting point) towards the cell (3,3) increases (cells (2,2), (3,2), and (4,2)), but the lateral and downstream cells (in the vicinity of the target) will remain unchanged. In this section in order to elucidate the issue, the example mentioned in the preceding section is taken into account, this time, the cell (3,3) is set as a barrier-containing cell.

Step A: At first, the value for the three cells upstream of the barrier, namely cells (2,2), (3,2), and (4,2) should be rectified. As reflected in Figure 30b, having said that cell (3,3) is the nearest cell to the target among their neighboring cells, due to the presence of a barrier, the position of this cell cannot be extrapolated and should be obviated from neighbor cells. Thereby, cell (2,3) is considered as the appropriate neighbor for the cell (2,2), and in the same vein, cell (4,3) would be the proper neighbor for cells (3,2) and (4,2). This valuation procedure is reviewed in Table 4, and its result is revealed in Figure 30c, according to which both cells (2,2) and (4,2) attain the least value and makes up the bedrock of pursuing the process at its next phase.

Step B: The adjacent cells of the two cells (2,2) and (4,2) must be identified and weighted in this step. This process has been performed in Table 4, based on which three cells (1,2), (3,2) and (5,2) acquire the lowest value, and will build the foundation to progress the process for the ongoing step.

Step C: Continuing the process at this step (Table 4) and evaluating the cells, neighboring cells with the lowest value are assigned to cells (2,1) and (4,1), both of which will make the cornerstone of the continuing process till the next step.

Step D: Valuation of neighboring cells at this step indicates that no cell holds the lowest value in contradistinction to the rest of cells (all the cells are of equal value) and consequently, the valuation process is completed.

The terminal grid elicited from this process illustrated in Figure 30d.

It is important to stress that the indicator positioned in the cell (3,2) will not prompt the robot to return to its beginning point for two reasons, one of which is that the optimal path has been left behind, and another one is that D^* algorithm will not become trapped in local minimum [30, 31, 33]. Furthermore, value modification at any given time leads to robust decision-making in such a way that there will be a likelihood of generating optimal paths during encountering obstacles.

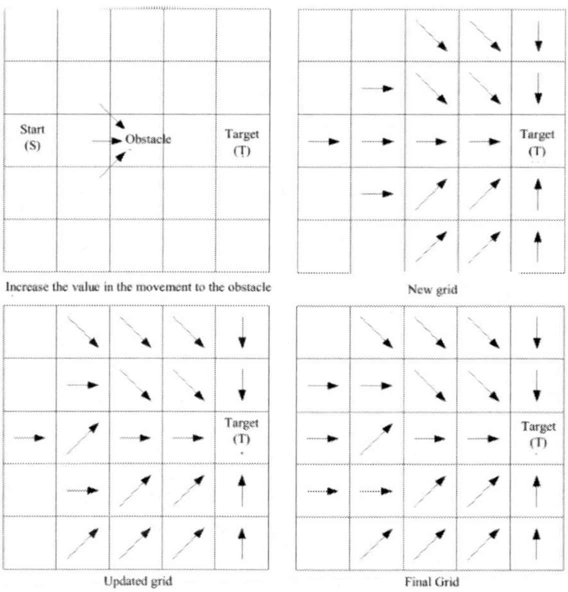

Figure 30. The process of moving vectors to a path in a grid with an obstacle.

Com Algorithm

Com algorithm is implemented in order to articulate how the robot quits the obstacle's boundary when the requirements of which are gratified.

Table 4. Evaluation of cells that contain the pointer into cell (3,3)

Valued cell	Neighbor cell with at least distance to target	Step value from (1) to (2)	Step value from (2) to (T)	Total value of cell
evaluation of cells that contain the pointer into cell (3,3) (Figure 28a)				
(2,2)	(3,2)	1	2.414	3.414
(2,3)	(3,4)	1.414	2.414	3.828
(2,4)	(3,4)	1	2.414	3.414
Table of flow frequency for the others cells by the algorithm (Figure 28b)				
(2,3)	(3,4)	1.414	2.414	3.828
(2,1)	(3,2)	1.414	2.414	3.828
(2,5)	(3,4)	1.414	2.414	3.828
(1,4)	(2,4)	1	3.414	4.414
(1,2)	(2,2)	1	3.414	4.414
(1,3)(S)	(2,2)	1.414	3.414	4.828
(1,5)	(2,4)	1.414	3.414	4.828
(1,1)	(2,2)	1.414	3.414	4.828
continuing the valuation flow of selection cells' neighbor cells from previous step				
(1,2)	(2,2)	1	3.414	4.414
(1,4)	(2,4)	1	3.414	4.414
(1,3)(S)	(2,3)	1	3.828	4.828
(1,5)	(2,4)	1.414	3.414	4.828
(1,1)	(2,2)	1.414	3.414	4.828
continuing the valuation flow and creation final grid (Figure 30d)				
(1,3)(S)	(2,3)	1	3.828	4.828
(1,5)	(2,4)	1.414	3.414	4.828
(1,1)	(2,2)	1.414	3.414	4.828

In this algorithm, the principles of leaving the barrier are evidently anticipated; ergo, they can be harnessed to develop Bug algorithms. Operating conditions of Com Algorithm are enumerated as follows:

Step 1. Take steps towards the target until one of two scenarios ensues as follows:

Step 1.1. The target is achieved, the algorithm is terminated successfully.

Step 1.2. An obstacle is faced, the barrier's boundary should be followed according Step 2.

Step 2. If the possibility of moving towards the target exists, move out of the obstacle's range, return to Step 1.

As stated above, the strategy based on which the barrier can be left behind is Com algorithm for this very reason, attainment of the target for the robot through this algorithm is not guaranteed. As an example, the implementation of this algorithm in an environment similar to Figure 31 illustrates that the robot cannot achieve its target and continually turns around the barrier's border.

Class 1 Algorithm

An alternative method to scrutinize how to go away from the obstacle is "class 1" Algorithm, which does not ensure that the robot reaches the target akin to Com Algorithm as its counterpart. Departing from the barrier in this algorithm is conducted at the closest point to the target with respect to other traversed points. The algorithm class 1 can be applied for the development of Bug algorithms based on its rules to overstep the obstacle analogous to Com algorithm. The performance principles of the algorithm class 1 are laid down as follows:

Step 1. Progress towards the target to meet up with one of the following scenarios:

Step 1.1. The target is achieved, a successful exit from the algorithm.

Step 1.2. An obstacle is come across, trace the obstacle's boundary, and go to Step 2.

Step 2. If the movement to the target is feasible, move away from the obstacle's area. Return to Step 1.

It should be pointed out that in Step 2, the nearest point for the robot to the target appears to be the leaving point as compared with all the points from which the robot has already passed. Execution of this algorithm in an optional environment is considered in Figure 32.

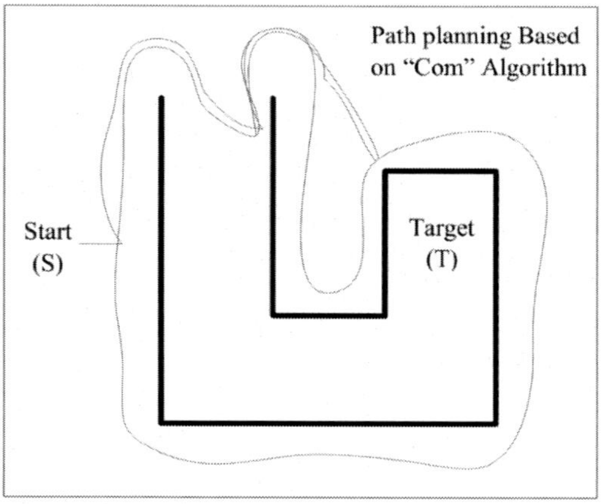

Figure 31. Implementation of Com algorithm in a hypothetical environment.

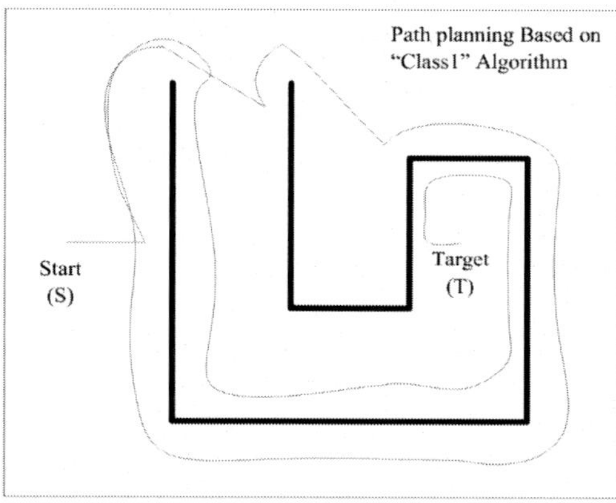

Figure 32. Implementation of Class 1 algorithm in a hypothetical environment.

Rev 1 and Rev 2 Algorithms

Algorithm "Rev 1" has been presented by Horiuchi and Noborio [56]. The algorithm is almost the same as ALG1, but the distinction between them is made based on frequently reverses in a direction following of the robot in each round in addition to planning the lists of *H* and *P* with the objective to record episodes more efficiently. The structure of the algorithm Rev 1 is composed of the following stages:

Step 1. Progress towards the target point until satisfying one of two conditions below:

Step 1.1. Attainment of the target is fulfilled, stop the algorithm; the mission is successfully performed.

Step 1.2. An unspecified obstacle has come across, the collision point H_i is determined and whose detail characteristics are recorded in order to modify the lists *H* and *P*.

Step 2. Movc in H_i and scan the list *H*. If both routes in H_i were checked out before, they would be expunged promptly from the recorded list *H*. Then the robot adheres to the barrier and undertakes to revolve around it until one of two scenarios is accomplished:

Step 2.1. Hit the target; the mission is successfully carried out.

Step 2.2. If the distance between the robot's current position and the target is less than the all distances of its past locations to the target (metric requirements) and moving directly towards the target is possible (physical requirements), the specifications of leaving point should be documented in the list *P* and returns to Step 1.

Step 2.3. If the robot encounters the previous collision point H_i, the mission has collapsed. This is the condition in which the barrier's boundary completely enshrouds the target.

Step 2.4. If the robot comes back to the collision point, the previous H_i is documented as the next Q_l in the recorded list *H*. Thereafter, the robot according to the information submitted in both *H* and *P* lists takes the shortest path to meet the collision point H_i and from this path moves towards this point. Meanwhile, the robot this time traces the border in the opposite direction.

Step 2.5. If the robot returns to a past leaving point, this point will be recorded as the following point in the list *H*. Subsequently, the shortest route for the achievement of the collision point H_i is selected. The robot then approaches this point through the information existed in the lists *H* and *P*. This once, the robot pursues the barrier's wall in the opposite direction.

Algorithm "Rev 2" was also founded by the same developers of "Rev 1" [52]. The dissimilarity between Rev 1 and Rev 2 lies in Step 2.5 in consequent of which if the robot in the algorithm Rev 2 return to the prior leaving point, and the amount of the next point in the list *H* would be the previous leaving point. After that, the robot based on the comprising information in the lists *H* and *P* selects the shortest path towards the point of collision H_i through which it proceeds towards this point. At this time, the robot follows the obstacle's border in a reverse direction. As an example, the representation of these algorithms' performances is displayed in Figure 33.

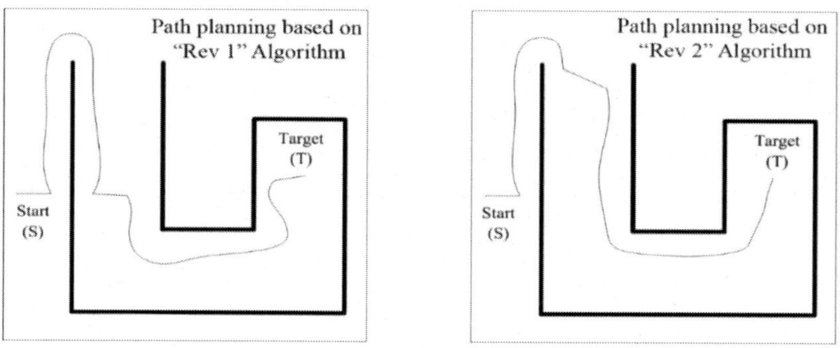

Figure 33. Implementation of Rev 1 and Rev 2 algorithms in a hypothetical environment.

Two Chosen Methodologies for Routing in Unknown Environments

Two specific methodologies in the area of routing in unknown environments are going to be presented in this section. The first selected method is the histogram of vector field that is elaborated by Borenstein and Koren [2]. The focus of this strategy is gridding the environment into a set of cells with the aid of sensors, and it then associates a vector to each cell of this grid. Ultimately, the integration of these vectors and the application of algorithms for obstacle circumvention lead to the selection of the appropriate path for the robot. The second strategy that has been evolved by Bianco and Cassinis [3] is termed multi-strategy. In this strategy, the robot implements particular algorithms, which are conducive to unknown environment, together with pre-determined maps from the environment.

Histograms of Vector Fields

In this section, the case in which the current position of the robot and the target positions are unknown is taken into account. Whatever the case, the robot is expected to be capable of detecting the target position. To do so, whether optical or audio sensors or imaging cameras may be utilized. In addition, algorithms identical to Bug 2 could be applied with the purpose of bypassing the obstacle. Vector Field Histogram (VFH) approach undertakes to grid the environment implementing sensors, and a vector is also assigned to every cell of the grid in this method. The main idea behind this approach is the combination of these vectors, which results in the determination of how the robot can reach to the target point, and on top of that, obstacles bypassing algorithms can be employed by the robot so as to approach towards the target.

A network of the two-dimensional environment is exemplified in Figure 34, which can be harnessed in order to generate the VFH. In practical terms, to take into consideration all the cells indubitably would be both a time-consuming and labor-intensive procedure. Consequently, a part of the whole environment is merely considered to reduce the complexities of the calculations; this simplification method is known as "active grid" (Figure 34). Selection of the active grid is occasionally made regarding the effective range. Furthermore, this grid tends to be ever-evolving with respect to the movements of the robot. This two-dimensional grid is exploited to construct a histogram that evaluates the angle and the speed with which the robot can be controlled.

Not only is the histogram computed just for the cells that exist in an active grid, but also it can be generated when the robot solely covers up the local information. A calculated vector for each cell embodied two attributes:

- Direction, that is oriented from the grid cell towards the position of the current cell within which the robot is situated.
- Magnitude, which is contingent on the transparency of obstacles' presence in that cell.

Vector's magnitude, which denotes the value of each cell, can serve as a controlling force requisite for the robot's keeping clear of the obstacles. As an illustration, a vector drawn in an active grid is depicted in Figure 35. As evidenced by this figure, the number attributed to each cell encodes the likelihood of obstacle occurrence in that cell, and vector's length is dependent on the magnitude of this value.

Calculation of Each Vector

Presume that (c_x^{ij}, c_y^{ij}) is the position of cell (i,j), value of which is c_{val}^{ij}, and (r_x, r_y) indicates the current position of the robot. The principal objective is to procure the vector v^{ij} of the cell c^{ij}. The method for calculation of the direction and magnitude of the vector v^{ij} is as follows:

$$v_{mag}^{ij} = {c_{val}^{ij}}^2 \left(\phi - \sigma\sqrt{\left(c_x^{ij} - r_x\right)^2 + \left(c_y^{ij} - r_y\right)^2}\right) \tag{3}$$

$$v_{dir}^{ij} = \tan^{-1}\left[(c_y^{ij} - r_y)/(c_x^{ij} - r_x)\right] \tag{4}$$

where v_{mag}^{ij} and v_{dir}^{ij} stand for the length and direction of the vector, respectively. Moreover, φ and σ are the normalization constants. Besides, r_x and r_y refer to the components of the current position of the robot in the Cartesian coordinate system, which is the center-to-center distance from the robot to the cell in two directions. It is imperative to emphasize that in this relation when $c_x^{ij} - r_x \geq 0$, the result should be sum up with 180.

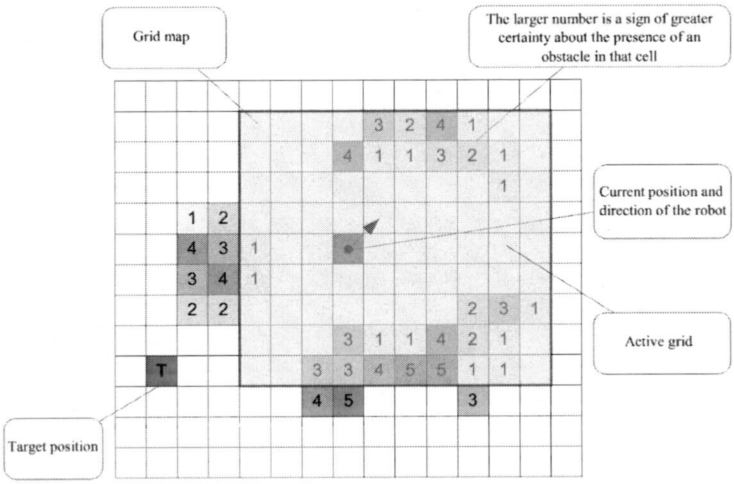

Figure 34. Gridded map in the histogram and its specifications.

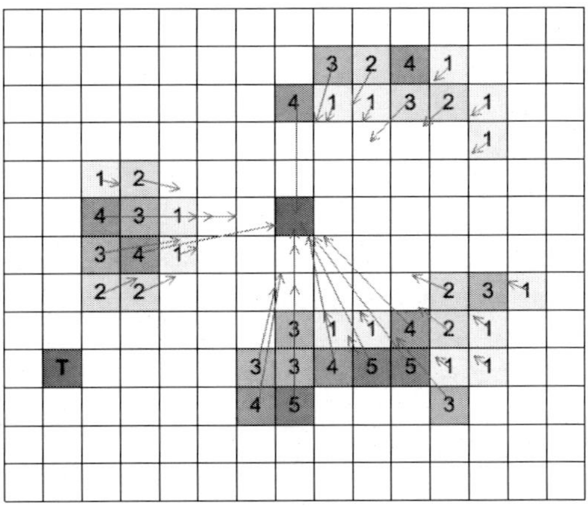

Figure 35. Calculated vectors for cells in an active network.

Calculation of Sectors

This section deals with the sector principle that can be applied to determine the extent of accumulation of vectors drawing from obstacles to the robot. Correspondingly, angular division α, which is the angle division available to the robot, is defined. In other words, the field of vision of the robot is divided into α sections, all of which are of the same amount. For instance, once the robot has a full field of view of 360 degrees if $\alpha = 24$, the size of each sector would be 15 degrees, and if $\alpha = 16$, the size of each sector changes into 22.5 degrees (Figure 36). The sector corresponding to each cell can be calculated as follows:

$$s^{ij} = int(v_{dir}^{ij})/(360/\alpha) \tag{5}$$

in which integral part function is denoted by "int".

Histogram Calculation

The number of α sectors should be taken into account for the histogram calculation. The value of each sector that is indicated by h_k and defined as "density of polar barrier" is equivalent to the sum of magnitudes of all vectors that have the same s^{ij}. Herein, h_k is calculated based on the following formula:

$$h_k = \sum_{i,j}(\boldsymbol{v}_{mag}^{ij} \,|s^{ij} = k)\ (0 \leq k \leq \alpha) \tag{6}$$

in which k refers to the under-study sector's number. Accordingly, each vector v^{ij} is assigned to its associated sector. An illustration of vector assignments to their associated sectors is depicted in Figure 37.

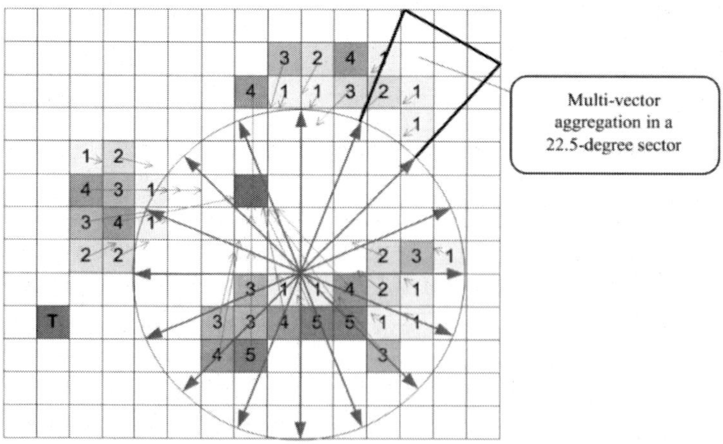

Figure 36. Division of the robot's field of view.

Histogram graph can be constructed by determining h_k for each sector. As shown in Figure 38, the histograms of the environment for two 16- and 72-sector grids are taken into consideration. As demonstrated in histogram graphs, the slots (having zero amount) in 16-sector case have lower numbers; considering the number of zeros for the 72-sector case is higher, this could be expounded that open paths for the robot to traverse is more. Put differently, the higher the number of sectors, the more the prospect of detecting an unobstructed path for the robot.

A novel idea for matching the histogram is replacement of one value from histogram with the average amount of before and after it. This concept is illustrated in Figure 39, according to which the original histogram (without averaging) has more dispersion. Once the average of the values of four nearest neighbors of a sector is worked out for one time, the resultant graph will turn into a more suitable form. Iteration of this process for a couple of times can contribute to the further improvement of the histogram. It must be emphasized that the averaging operation should proceed to such extent that border amount ε which is going to be explicated in the subsequent section is fulfilled.

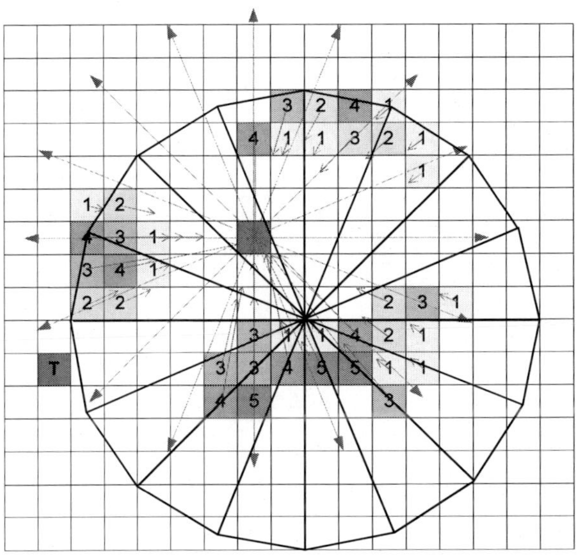

Figure 37. An example of assignment of each vector to its own sector.

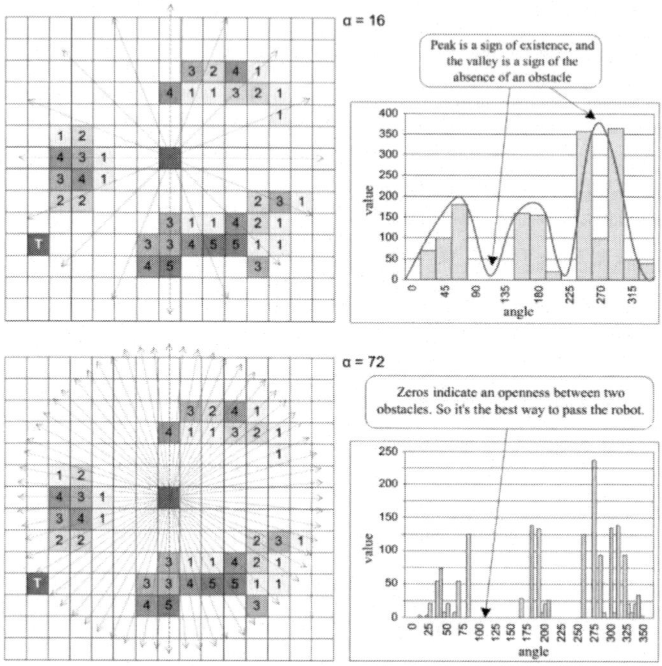

Figure 38. The histogram of the sample environment and its dependence on the number of sectors.

How to Use Histogram

At the outset of the histogram generating process, it is of paramount importance to specify the sector comprising a vector that is directed from the robot position to the target point. Towards this end, the following relation is established:

$$s^g = int\left[\tan^{-1}\frac{g_y - r_y}{g_x - r_x} \Big/ \frac{360}{\alpha}\right] \tag{7}$$

wherein s^g indicates the sector containing the target vector, and (g_x, g_y) denotes the coordinate of the target point. By drawing a histogram, the valley, which is as close as possible to the sector, s^g would be retrieved, among the sectors $s^i, s^{i+1}, \cdots, s^{i+k}$ where i is the first obstacle-free sector, and k specifies as the number of barrier-free sectors. As a case in point, the target position in the histogram grid exhibited in Figure 40, is associated with the sector from 210 to 215 degrees. The position of the under-consideration sector in this histogram is also displayed in Figure 41 along the correspondent column. The nearest valley to the assigned sector that is typically signified by s^c is positioned at the location 220 through 225 degrees and is delineated with a circle on the histogram. More importantly, it is compulsory that the size of all sectors of a valley be lower than border amount (ε) (Figure 42).

The algorithm aims to compute the direction of the robot's movement; therefore, a certain number of criteria is needed to be satisfied:

- The number of sectors in a valley which is expressed with k.
- The positions of border sectors which are symbolized by s^i and s^{i+k}.
- The position of the nearest sector of a valley to the target sector, which is identified by s^c.
- The number of neighboring sectors s^c which is represented with ±max (max is a user-defined value)
- The position of border sectors neighboring s^c is specified with $s^f = s^{s \pm max}$.
- The border value ε which is determined the maximum size of the sectors inside a valley (ε is a user-defined value).

Following these criteria, valleys can be categorized into two classifications:

- Vast valley where $k > max$
- Narrow valley where $k \leq max$

An example of a vast valley with $k = 4$ when $max = 2$ is presented in Figure 43. Now, assume that the sector s^c, among all the sectors $s^i, s^{i+1}, \cdots, s^{i+k}$, is closer to s^g. After quantifying of s^f, the robot can initiate its movement along $\theta = (s^f + s^c)/2$. As seen in Figure 43, the robot tends to progress along the sector s^{i+1}. It should also be pointed out that the higher border value ε, the greater the speed of the robot movement, and in turn, highly improbable to confront obstacles. Nevertheless, the robot is confused in extra-large values of ε.

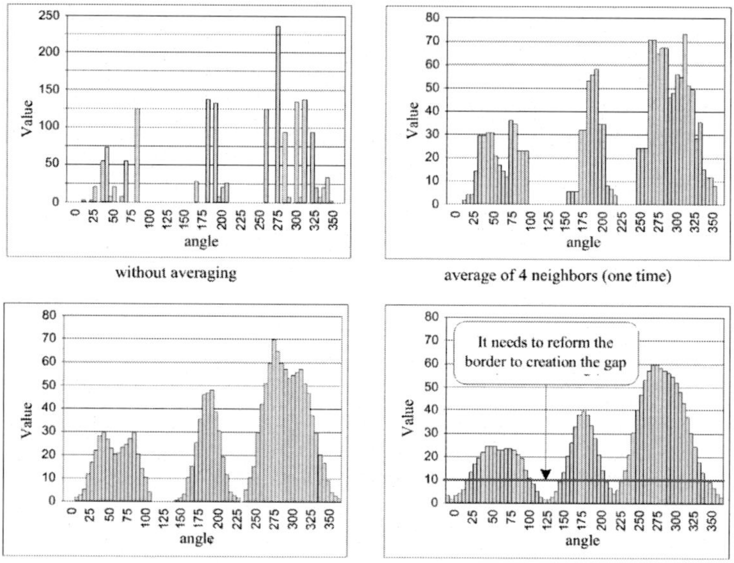

Figure 39. The histogram's dependency on the frequency of averaging the neighbors.

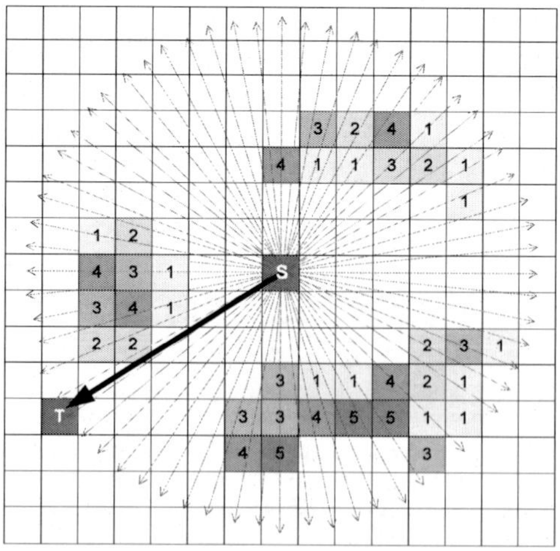

Figure 40. Specifying the sector including the vector of the current position of the robot to the target position.

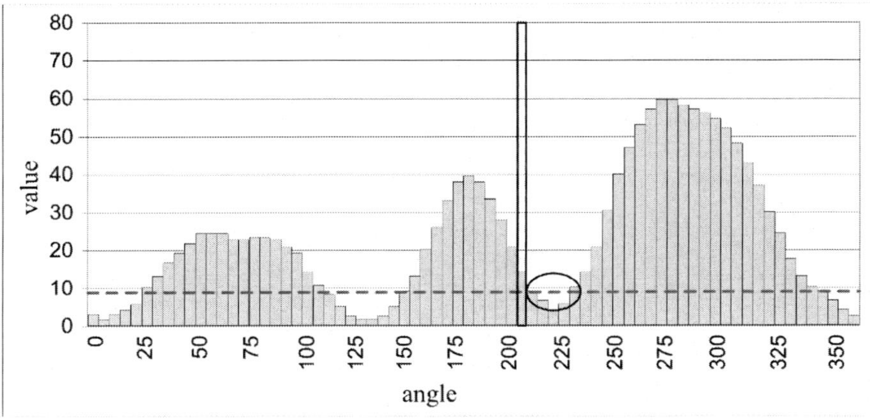

Figure 41. Determining the nearest valley to the sector s^g.

Designing Path Based on Multi-Strategy Approach

In this section, a design for routing robots in an uncertain environment which is equipped with the independent control system is outlined in such a way that multiple strategies towards reaching their target position is adopted. This approach has the capability of learning that enables the robot to utilize the outcomes of foregoing experiments and make the most of them to carry out better maneuver in the future.

As already mentioned, robot guidance in an unknown environment has two resolution options:

1. Providing the robot with a pre-specified map from the environment that has the capacity for being observed from the starting point by a robot.
2. Implementing distinct algorithms that are efficient enough to cope with unknown environments.

This section is dedicated to the consideration of the assortment of these two solutions with an eye on the integration of their advantages and exceeding many limitations they impose. That is to say; it indicates that recalling some features of the operational environment, in conjunction with the application of algorithms, which are robust to environmental variations, would be functionally beneficial.

The proposed scheme with the aim of designing the robot path makes use of two primary models founded on the two diverse hypotheses. The first model, chosen from Ref. [21], is established upon information from the environment. Based on this model, numerous designs have been developed the best of which are Lozano-Perez [57] and Brooks [5]. On the other hand, the second model attempts to design the robot route with the initial partial information from the environment. Incomplete data can be steamed from sensor-related reasons. Given the fact that it is unfeasible to regard the environment comprehensively right at the beginning of the process, path designing is divided into several time intervals [17].

Recommended strategies towards the environments, which are unknown from the very start, contribute to the development of path planning algorithms; however, it will not inevitably result in well-converged and optimal paths. Having this in mind, some researchers have presented a variety of techniques with the purpose to promote convergent and optimum strategies. As an example, Bug 2 Algorithm as the first strategy that ensures convergence to target was executed, and if converging was doomed to failure for any reason, Bug 1 Algorithm could be applied [16]. Path planning through intelligent training techniques was also developed by several investigators in such a way that re-planned paths can access to the information of previous paths successfully reached the target [58].

Concepts and Definitions Used in the Proposed Multi Strategic Method

In this section, path planning implementing different types of multi-strategic approaches is going to be enunciated in which not only is a failure (such as divergence) of one strategy in a number of environments taken into consideration, but also many aspects of constraints including the kind of under-used sensors are concentrated. It is important to note that, failures (loops and more broadly, divergences) are by no means the only peculiarities giving rise to inefficacies of a specific strategy in some environments. Actually, the assessment of the surrounding environment for the robot is prone to many limitations. Take a strategy incorporating acoustic sensors as an example. Even supposing the well-functioning of these sensors in an open environment, they would become inefficient in narrow and tight spaces.

In order that a principle for this concept is formulated, the impact factor function η and two sets S and C are devised, definitions of which are provided as follows:

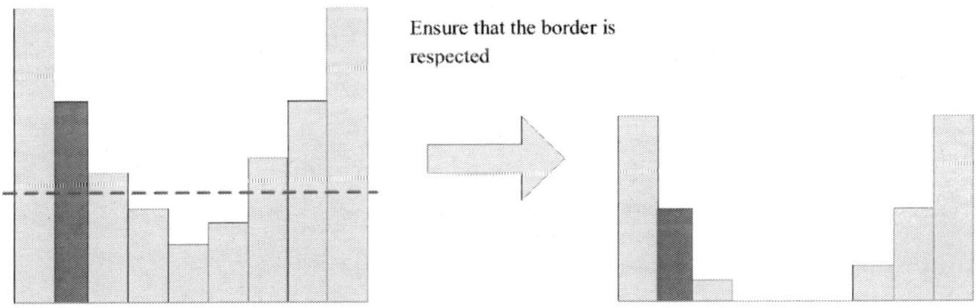

Figure 42. Cutting the histogram regarding the boundary value ε and finding the sectors of a valley.

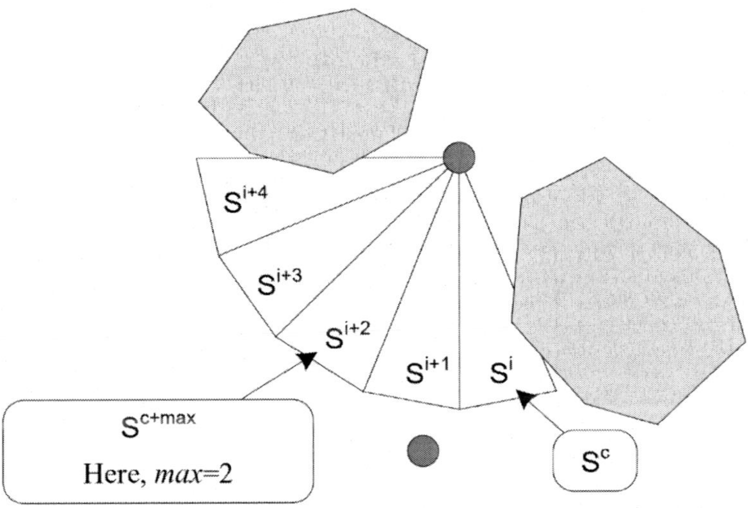

Figure 43. An example of a vast valley ($k = 4$, $max = 2$ and $k > max$).

Set *S*, elements of which are possible strategies in path designing and are displayed as $s_1, s_2, \dots, s_n$;

Set *C*, elements of which are possible environmental characteristics;

Function η, which denotes the associative impact factor of a strategy recognized in traversing a given path *P* and value of which lies within the range of $[0,1]$.

If the deliberative strategy is bound to fail on the given path *P* (meaning that either the target is not achieved or this path is a loop), η is presumed to be zero. If so, the path to obtain the target is considered optimal (the optimal path with two criteria of the shortest route and maximum possible speed is defined), and η will take the value of 1. Otherwise, in cases where ($0< \eta <1$), the specified strategy meets the target on all occasions, but the traveled path cannot fulfill the optimal route conditions. If $s_n(c)$ is the optimal path $(s_n \in S \; and \; c \in S)$, this assumption can be evaluated as the following relationship:

$$\forall s_n \in \Rightarrow \exists \overline{c} \in C, s_m \in S{:}\, \eta\big(s_m(\overline{c})\big) > \eta(s_n(\overline{c})) \tag{8}$$

where $\overline{c}$ refers to a particular environmental element appertaining to "c"'s, it would be inferred that an environmental parameter for each strategy could be found so that it stimulates its highly efficient performance contrasting to the current strategy. Stated differently, no strategy could supposedly be definitely optimal.

In this context, the adopted method for division of the environment is its conversion into square pieces, dimensions of which are approximate of equal size to the robot dimensions. This procedure has been vividly portrayed in Figure 44.

The full path from the start point towards the target position is a sequence of cells (or boxes) that are not necessarily adjacent to each other ($A_5 \rightarrow A_1 \rightarrow \cdots \rightarrow A_n \rightarrow A_T$). Thereby, the sequence of consecutive paths, which are shown with $A_i \rightarrow A_j$, constituted the cumulative path. Ergo, it would be feasible for routes to be planned pursuant to various strategies.

Selecting an appropriate strategy to design each route is directly dependent on the robot itself. The robot performs this task with reference to a body of information that is stored in terms of features belonging to each strategy. These features, which make it possible to draw a comparison between different strategies possible, are pointed out as follows:

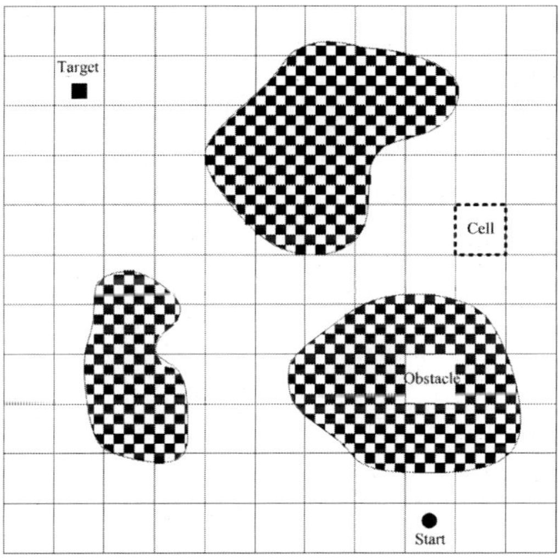

Figure 44. The division of the environment into cells.

- Usability in a particular local environment,
- Potential efficiency of path traversal,
- The actual efficiency of path traversal.

Establishing usability criteria is extremely complicated for a specific strategy. Conversely, unsuitability can be defined as "for any given strategy s_n, there is at least one environmental parameter s_{ij} based on which strategy s_n is guaranteed to be failed." This means:

$$\eta\left(s_n(c_{ij})\right) = 0 \tag{9}$$

in this equation, c_{ij}, which is a member of the effective environmental parameters, is designed employing strategy s_n (sequence of consecutive paths starts from a cell A_i to cell A_j). In order to develop deeper comprehension, some critical points related to impact factor function η are presented.

Presumable impact factor: This factor is defined as the ratio of minimum navigation time (t_{μ}) to presumable navigation time (t_{σ}). Accordingly, after determining both minimum time $(t_{\mu ij})$ and possible time $(t_{\sigma ij})$ for any given path $A_i \rightarrow A_j$, presumable impact factor can be derived from the following equation:

$$\eta_p = \frac{t_{\mu ij}}{t_{\sigma ij}} \tag{10}$$

Minimum navigation time refers to the period during which the robot navigates a path directed from starting point (placed at the cell A_i) to the target point (placed at the cell A_j) with maximum speed. By the same token, the presumable time indicates the period required for the robot to meet the target point through a specific strategy.

Actual impact factor: This factor is defined as the ratio of minimum navigation time (t_{μ}) to actual navigation time (t_{μ}), and for any path $A_i \rightarrow A_j$ from the following relation is procured:

$$\eta_p = \frac{t_{\mu ij}}{t_{\eta ij}} \tag{11}$$

in this relationship, $t_{\eta ij}$ indicates actual navigation time, during which the robot moves across a path. Indeed, presumable time can be computed in advance of design step; while the actual time should be specified only after robot locates at the target point.

With the aid of environmental information, the robot can reach a decision on the most efficient strategy amongst recorded strategies in its memory in order to proceeds from cell A_i to another cell A_j. In this way, for any recognized practical strategy $(\eta(s_n(c_{ij})) \neq 0)$, the robot decides to choose a strategy with the highest actual impact factor. If such a strategy is non-existent, a strategy with the highest presumable impact factor is selected.

Building an Environmental Representation Matrix

Selection in a multi-strategic approach is carried out according to the environmental presentation information by the robot. Environmental presentation defines as a square matrix, rows, and columns of which demonstrate the identity of cells, and every component of which is attributed to a table containing a complete list of all the so-called strategies. The robot should be accommodated with three items of information from the local environment, which are provided by three virtual sensors:

Virtual Sensor Applicability (VSA): A binary response (zero or one), which indicates the applicability of a specified strategy in a localized environment is provided using this sensor.

Virtual Sensor Presumable Efficiency (VSPE): A number in the interval $[0,1]$ based on the presumable impact factor $(\eta_p = t_\mu/t_\sigma)$ is computed using this sensor.

Virtual Sensor Actual Efficiency (VSAE): Akin to presumable efficiency but this time based on the actual impact factor relation $(\eta_a = t_\mu/t_\eta)$, a number in the interval [0, 1] is also calculated using this sensor.

These virtual sensors are executed with either general information from the environment or implementing active sensors (detailed explanation for integrating active sensors in path designing approaches are accounted by Lumelsky and Skewis [12]. Data extracted from active sensors and environmental information are utilized on the grounds that identifying environmental factors has demonstrated enormous potential for adversary affecting the usability of a strategy.

Towards this end, t_σ is quantified at first; this factor is generally dependent on the maximum amount of time spent by the robot to acquire information from the environment. The two primary methods for this are either applying mathematical relationships (to compute the time necessitated for receiving information at each time in a particular environmental setting) or making use of statistical information from employment history. Either way, VSA is A coherent framework of principles gleaned from human experimentation.

Matrix description of the route is demonstrated in Figure 45, according to which a table for each element of the matrix is designed such that it is comprised of usability, actual and presumable impact factors.

Path Design Methods Based on Multi Strategic Approaches

Path planning that is built upon multi-strategic approach contingent on the aggregate data from sensors collected from the surrounding environment can be accomplished by means of three methods, each of which has its own complexity and efficiency as follows:

Minimal Method

The robot designs its entire route premised on only one strategy until it leads to failure. The robot would employ other strategies in order to re-planning and sensors to locate strategies failure.

Median Approach

The robot alters its strategy once identifying an environmental condition that is impediment to the execution of its current strategy. Local environmental factors for the robot can be detected with the aid of sensors.

Full Method

In this method, the robot substitutes henceforward its current strategy with a strategy that has better performance in any local environment. The process implemented in each cell to evaluate the efficiency of the robot strategy functions based on receiving information from the sensors.

Here, the minimal approach is considered as the default approach. This way, initial information is preserved in each element of the matrix. When the robot passes from a specific element, the stored information can be enforced. Consequently, it is needless to compute all the information once again. The successes and failures in the implementation of any given strategy can be recorded, and by the time of seeking solutions for the following path planning problems in the same environment can be utilized optimally.

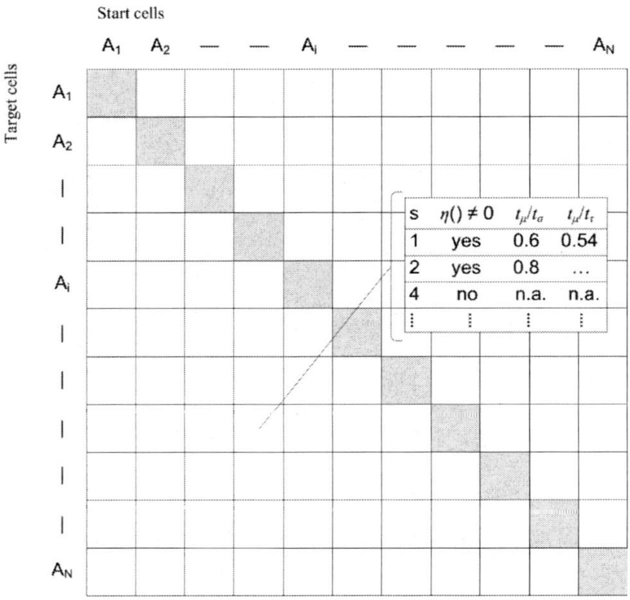

Figure 45. Matrix display of the path.

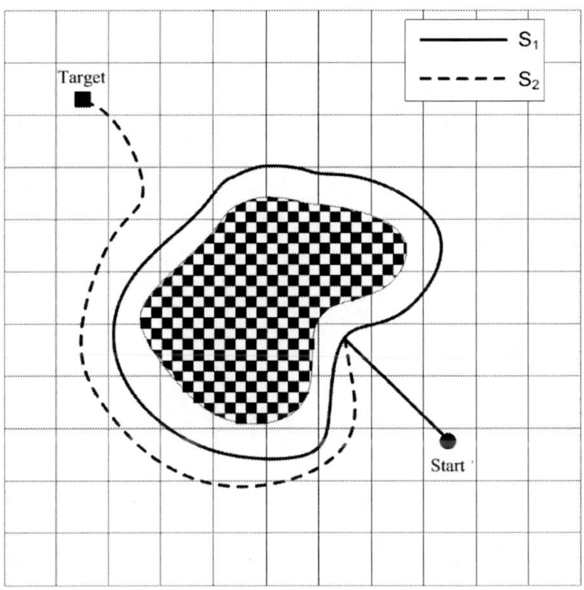

Figure 46. Change of strategy in case of being placed in a loop.

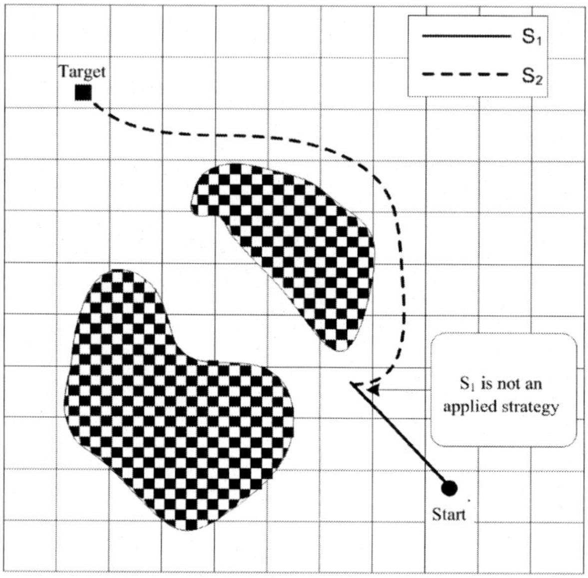

Figure 47. Changing the strategy by breaking it.

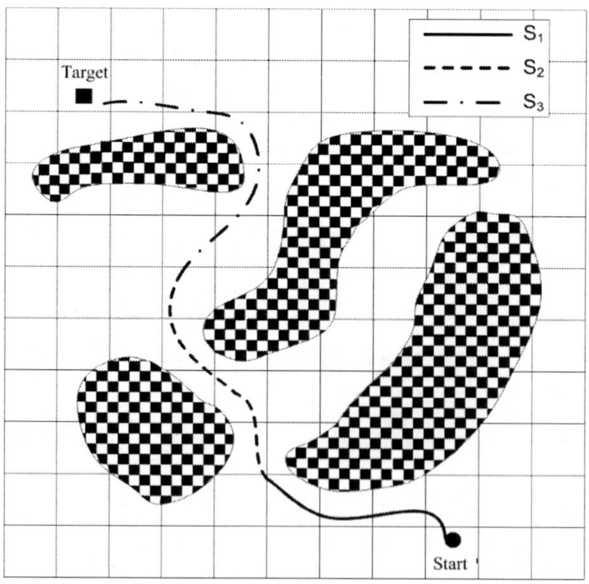

Figure 48. Applying the best strategy.

Three examples for executing the multi-strategic approach are presented from Figure 46 to Figure 48. In Figure 46, the strategy s_1 is implemented until failure takes place (the robot trapped in an endless loop). Afterward, the robot applies a replacement strategy (s_2). In Figure 47, strategy s_1 until failure occurrence (in this situation means discontinuity), and alternatively another strategy (s_2) is utilized. In Figure 48, strategy s_1 is executed as long as maintaining its supremacy over other strategies. Once superiority over another strategy is lost, the robot takes a strategy with greater efficiency (s_2 at first and then s_3). In Figure 48, the union of strategies s_2 and s_3 demonstrates improved performance in comparison with the pairs s_1 and s_2 strategies.

Output of Proposed Multiple Strategy Approach

The multi-strategic method is by no means a path planning “strategy”; it is quite a refined approach to design a path. As a matter of fact, implementation of current strategies sometimes is able to boost the efficiency of the entire route or tackle the limitations of a specific design strategy owing to the particular environmental circumstances that give rise to the impossibility of achieving the target position for the robot.

It is indispensable to provide a comparison between multi-strategic method and different enhanced design approaches instead of just comparing with some path planning strategies. From this perspective, the main distinction between multi-strategic method and its contemporary classical counterparts lies in the fact that each cell, which is not necessarily incorporated in the entire path, has its strategy to reach the target. It is the exact reason for which intelligent training in classic approach and behavioral training in multi-strategic methods differ with each other. Behavioral instruction empowers the robot to take qualitative in lieu of quantitative information about the local environment into account. That being so, the multi-strategic methods, altogether, is less influenced by the environmental alterations along with the movement of obstacles as compared to their classical equivalent.

The multi-strategy approach is extraordinarily fruitful if the environment undergoes minor changes. For further elaboration, take an unknown environment within which the robot would operate into consideration. This environment can be classified into either static (without modification) or dynamic (with changes) types.

In order to establish a relation for these two groups of environments, N can be represented as the number of cells being in the dividend environment E with the condition $C_{E_t} \subset C$ at time t. If t_1 and t_2 ($t_2 > t_1$) allude to moments at which the robot is required to design a route and M indicates the number of cells whose variables are $C_{E_{t_2}}$ with regard to $C_{E_{t_1}}$ as the conditions of the under consideration environment, qualifications of different types of environment can be formulated as follows:

- When $M = 0$, the environment is considered static;
- when $M = N$, the environment is considered fully dynamic;
- When $0 < N < M$, the environment is considered partially dynamic.

On this account, the environmental modifications can be bridged to the environmental representation. As requisite information of each cell in the multiple-strategy approach is collected, the multi-strategy approach in terms of behavioral concepts can be evaluated in a diversity of environments from static to fully dynamic as follows:

Static Environments

It is pointless to exert effort in order to enhance the performance of designing the environment through the multiple-strategy approach in this kind of environment. Contrarily, other methods, which are inspired by developing geometry maps from the environment, can generally yield better results.

Full Dynamic Environment

Due to the fact that the environment totally alters every time during which the robot moves across the environment with respect to its prior state, storage of environmental information is unprofitable. In this type of environments, coping with unknown environments using a multi-strategic method can be accomplished. Nevertheless, the quality of the training system has completely sacrificed.

Partial Dynamic Environment

Thanks to the fact that all parts of the environment never change; therefore, experiments which were conducted earlier are considered to be advantageous for designing of the subsequent paths in this environment.

Multi Strategic in Action

Taking advantage of the multi-strategic method provides a dynamic designing route which functions well in environments with subtle changes. In each of the three aforementioned techniques, namely minimum, median, and full approaches, a set of impact factors (e) for every cell, are determined as below:

$$PE = \{e_i, e_2, \cdots, e_n\}$$

The expanded conversion of this equation with statistical analysis is written as follows:

$$D = \{e_i * V, e_2 * V, \cdots, e_n * V\}$$

in which V stands for the numbers of times that the robot traverses from the correspondent cells.

Corresponding to the employment of D in line with a randomly selected number $i(i \in \{1,2, \dots, n\})$, strategy s_i can be formulated. In some cases, the robot may apply a strategy that is not evidently optimal. This selection opens up an excellent opportunity for the robot to initially choose its direction when sensors have already collected a snippet of information from the environment, and it has led to the large extent to which the environment is unknown.

Chapter 3

PATH PLANNING IN KNOWN ENVIRONMENTS

INTRODUCTION

This chapter comes to grips with the path planning problem in an environment with a prespecified obstacle. In this connection, the robot avail itself of the provided maps form its information or external resources in order to prepare maps of its actual environment. In the case of the former, the robot may be allowed to profit from either satellite images or a predetermined fixed map at its disposal. In this particular case, it can be expected that the specification of positions of both the robot and the target have met in advance. Consider the example of a GPS receiver which is capable of identifying the desired positions. There exists an issue of prominent importance, and that is that current algorithms have a heavy dependence on the type of map. With respect to this issue and offered locations of barriers, three strategies for the robot 's navigation towards the target are recommended. With the help of these techniques, the robot acquires comprehensive knowledge about all the obstacles present in the environment and can access a series of pre-calculated data for its guidance towards the target. Three primary guidance methods at issue are categorized as follows:

- Feature-based guidance
- Guidance with potential field
- Vector Field Histograms

In the previous chapter, vector field histogram strategy, which is weighed up as a conventional approach in both known and unknown environments, was described at length. This chapter proceeds to present the two remained procedures.

FEATURE-BASED GUIDANCE

Feature-based guidance is formed based on the structure of features maps and drawn its correspondent graph. Guidance towards the target is premised on an approach adopted to the graph. In this section, stages of the process, which includes features maps preparation, drawing the graphs and its consummation when the robot meets the target point, are represented in detail.

Preparing Feature Maps

The first challenge posed in the robot guidance reliance on features is that the robot has to traverse obstacles boundaries all the way. Several approaches have been proposed for this purpose. One example is, according to Figure 49, if all the borders of an obstacle are of different size, the robot can seek out some singular corners that it has bypassed before in order for its search be brought to an end. As depicted in Figure 50, the robot is able to get the most out of an indicator (suchlike a circle) via marking its searching starting point until it ferrets out its prior indicator again, ensuring that the obstacle is fully crossed over. Needless to say, this method will result in failure in those instances where the path has a narrow width. In other words, to put an indicator in a narrow passage (Figure 50) leads the robot to spot the sign during retroverting across the path so that it will discontinue its movement; however, it hasn't completely traversed the obstacle yet. An alternative method to guarantee that passing over the obstacle is successfully accomplished is based on considering the maximum perimeter of the barrier and also halting the robot once it passed by the obstacle. This procedure has been demonstrated in Figure 51.

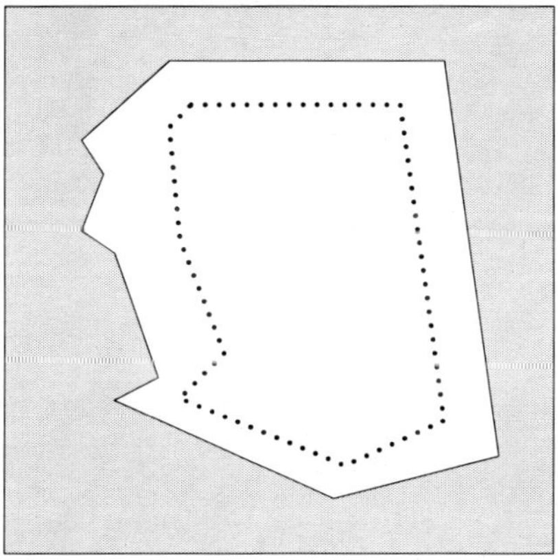

Figure 49. An obstacle with various kinds of edges; permitting robot's transverse with respect the edges.

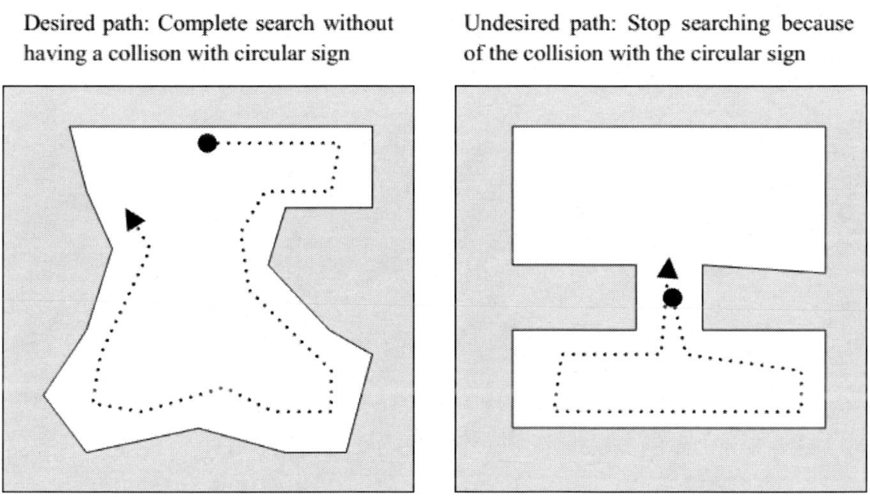

Figure 50. The result of the desirable and undesirable situations owing to implementing a circle markerr to traverse barriers.

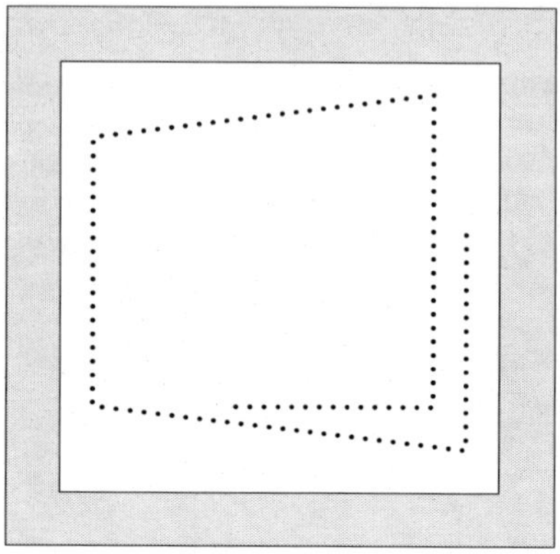

Figure 51. Suspension of robot's movement, having travelled the maximum perimeter.

Storing Feature Maps

There exist two strategies in order to store Feature Maps. In the first method, as illustrated in Figure 52a, during the course of bypassing every corner and edge, the robot stores their correct number in its memory. In the second method as shown in Figure 52b, a graph or a grid, which may be taken into account as a process analogous to neural networks, is represented for each individual object one by one.

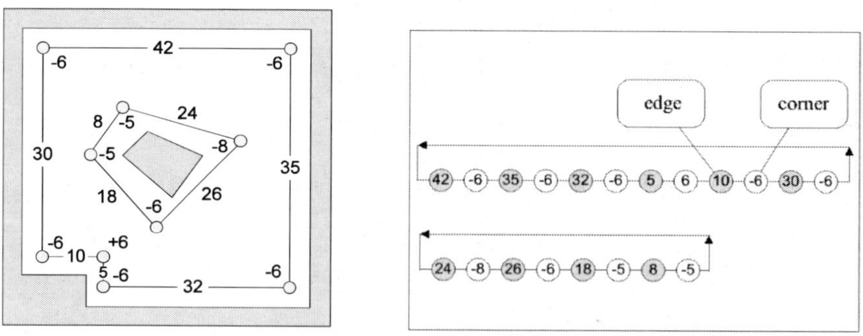

(a) Assign integer to corners and edges.

(b) Draw a graph correspondent to the environment.

Figure 52. Two diverse methods to record the feature maps.

Coinciding Edges

This section commences with making a comprehensive assessment of coinciding corners with one another. First and foremost, the two aforementioned distinct approaches to pursue the same exact environment are considered as depicted in Figure 53.

The task starts with questing for the first corner through following the straightforward procedure to deceit the new one. The border is selected for the coincidence's starting point. In this study, all the angles and edges are incrementally measured until either fulfillment of full coincidence is gained or otherwise, failure experienced (Figure 54). Whenever this evaluation is performed, one of these two states seem to be possible:

- Following the new one, contributes to the creation of new edges and corners.
- Following the new one, does not recognize existing distributions in the main map.

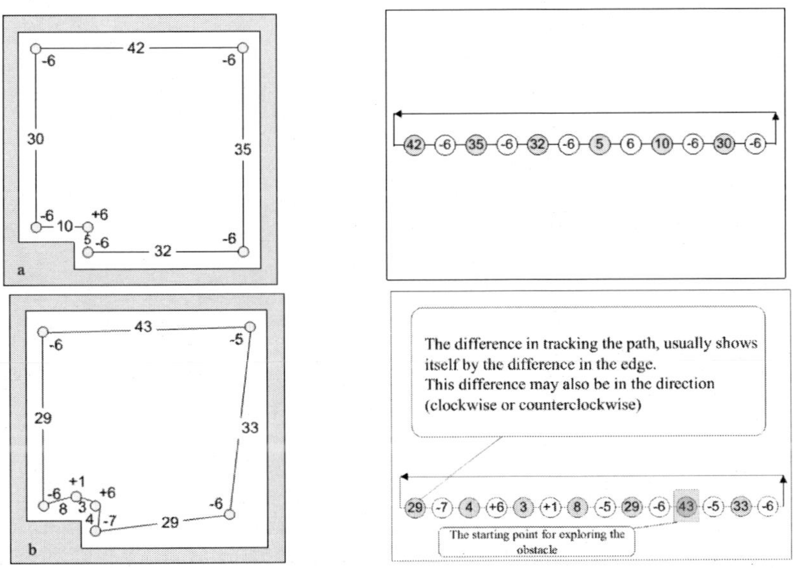

Figure 53. Two varied approaches to pursue a same environment and its correspondent graph.

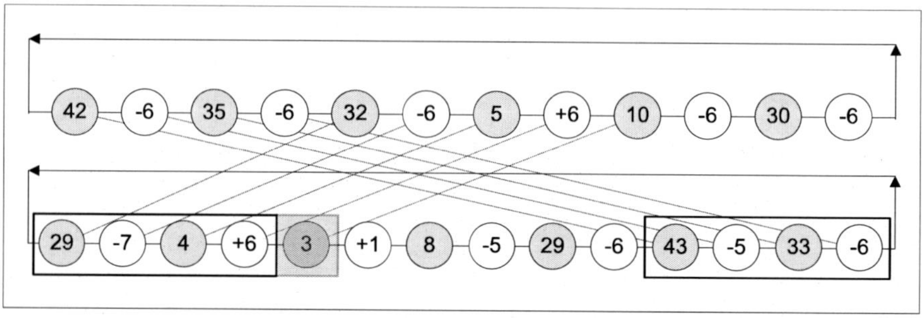

Figure 54. Evaluation of all angles and edges for coinciding.

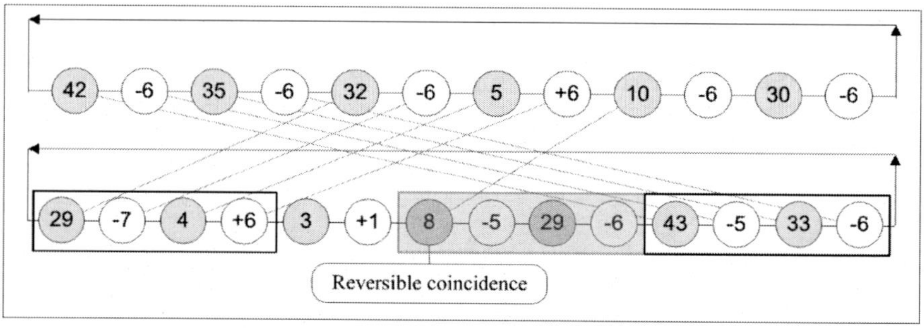

Figure 55. Coinciding edge 10.

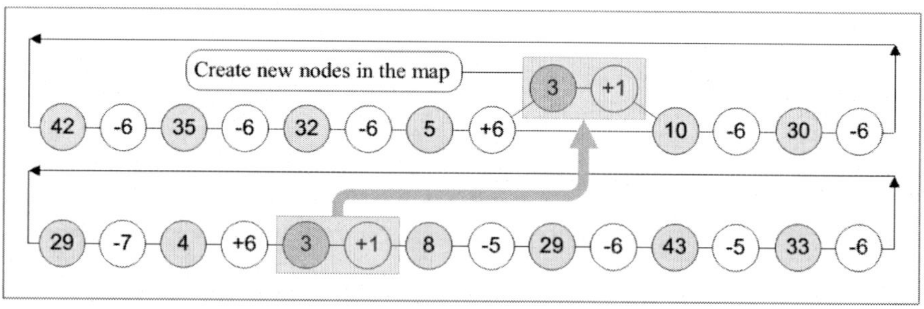

Figure 56. Introduction of the new edges.

In the former case, a coincidence for edge number 10 should be found regardless of the edge number 3 (Figure 55). In this step, new edges are formed into the map (Figure 56). Nevertheless, in the latter case, a coincidence for the edge number 3 should be sought (Figure 57).

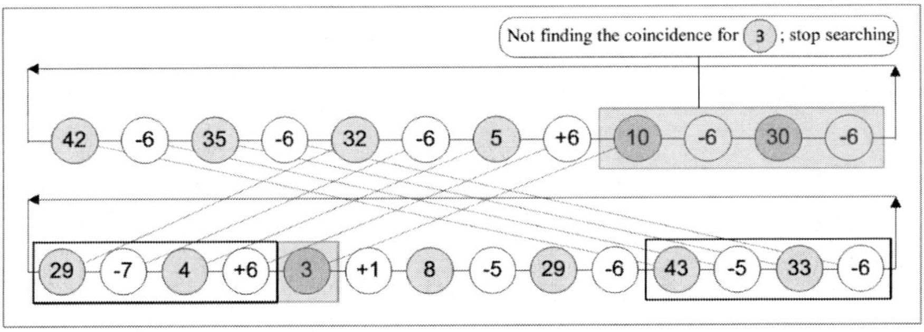

Figure 57. Coinciding edge 3.

Adding Bridges

In the course of drawing graphs, bridges can be taken into consideration as connecting edges to the obstacle (Figure 58). In a case where there is no required connection between two obstacles, it is of necessity to attach absolute numbers of bridges linking them that is conducive to moving across all the barriers. As demonstrated in Figure 59, before a bridge was inserted between these two obstacles, there was not any connections between their graphs, and only after two bridges are added, the connection between the graphs is established as well.

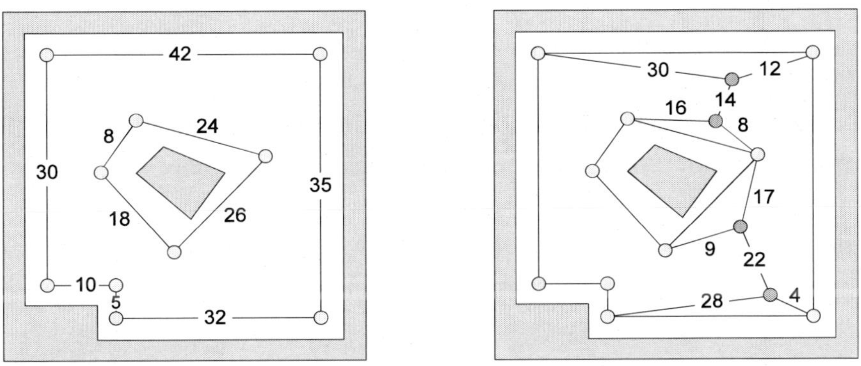

(a) before adding the bridge (b) after adding the bridge

Figure 58. An environment including convex and concave barriers, before and after adding bridges.

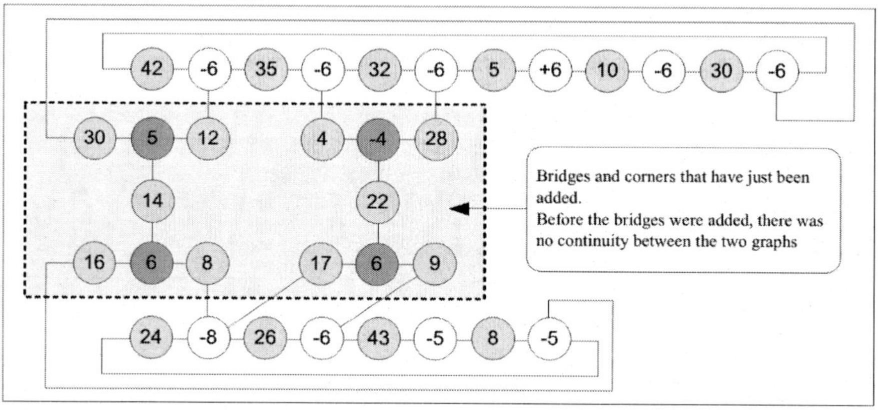

Figure 59. The environment's graph comprising of two barriers with added the bridges.

Graph-Based Guidance towards the Target

Guidance of the robot towards the target will be satisfactorily performed after creation of feature map by virtue of looking for the graph. One way of guiding the robot from its current position to the target position can be to seek throughout the graph, and then check whether the facing edge has a shorter length or not to identify a proper path. Thereafter, the robot traverses the obstacle in either a clockwise or counterclockwise fashion (Figures 60 and 61). To fulfill this objective, the position of one object within the graph is set to zero as a starting point; the movement of the robot starts from the determined point. While traversing each vertex, the value of the stepping is lengthened with the addition of the edge length of obstacle circumscribed between two vertices. This manner is introduced in Figure 62.

In this situation, the procedure through which the path of the robot can be selected should be determined. Equivalently, the decision is made whether the robot has to turn to its left or right once being located at a vertex. In order to do so, a sign is assigned to each edge; a plus mark for the edges located on the right and likewise minus sign for the edges on the left side of the robot. It is important to emphasize that whatever the case is, as part of the value setting process for the graph, absolute values should be attributed to the edges (Figures 63 and 64).

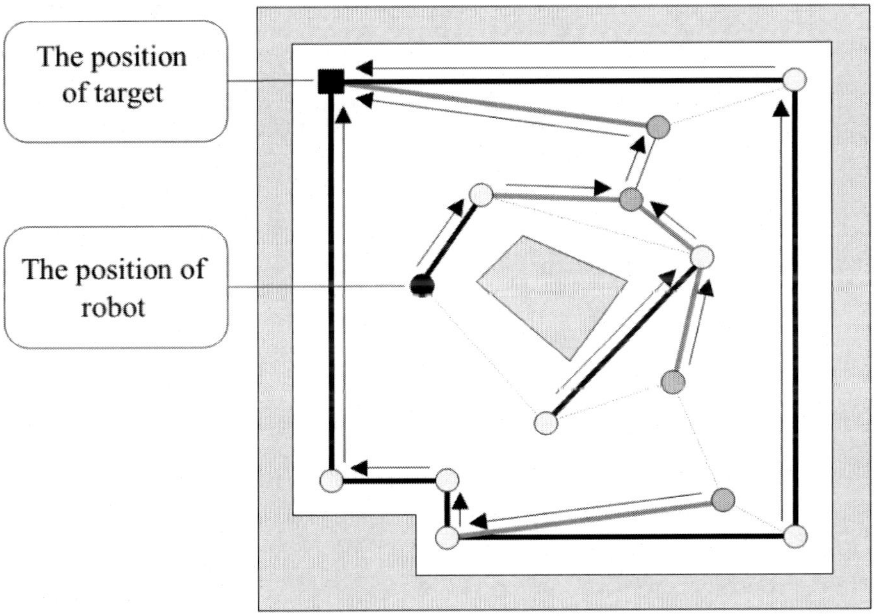

Figure 60. The environment's feature map including two barriers, guided utilizing guidance map.

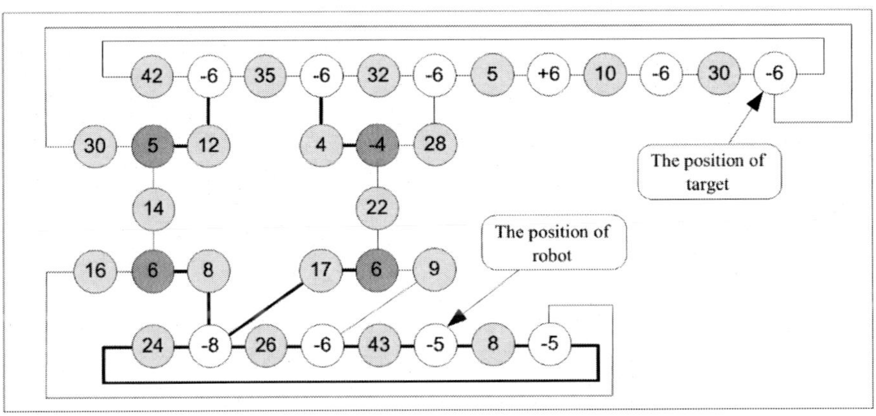

Figure 61. Graph of the feature map displayed in Figure 60.

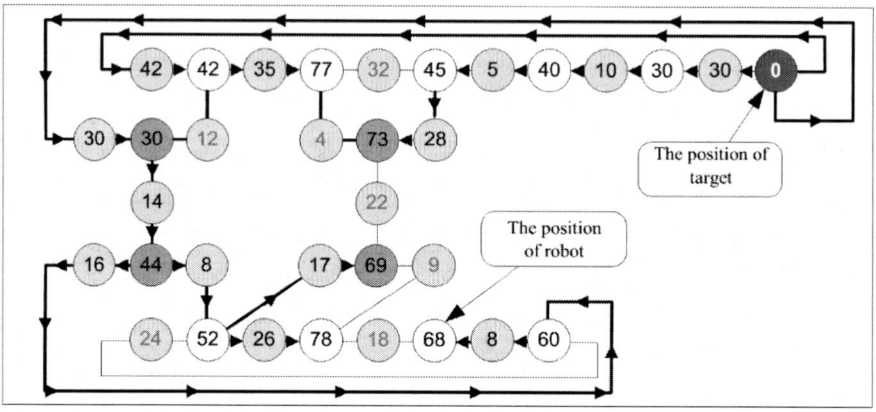

Figure 62. Guidance graph of Figures 60 and 61.

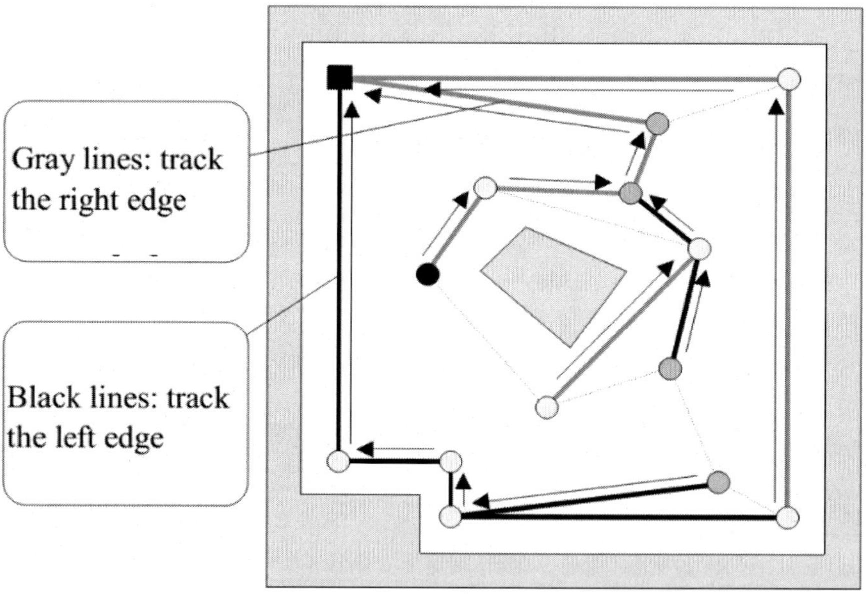

Figure 63. Feature map's graph.

Guidance in Potential Field

Guidance of the robot in the potential field must be conducted in conformity with the principle that the robot in the course of its motion in different environments can be either attracted or repelled by objects and situations, this can be thought of as behaving like a magnet. Robot as a means to be guided towards the target performs vector computation that is a function the obstacles' position and the location of terminal destination (the desired target point). The robot keeps moving along this vector until it is reached to the target position, thereby the calculated vector at each moment will also be updated. The principle of changing the robot path in the potential field is laid down based on the following concepts:

1. Attraction to the target point
2. Repulsion from the reference point (to avoid if the robot is standstill at this position)
3. Repulsion from obstacles

These three underlying principles are illustrated in Figure 65.

Each vector in the potential field has two main features that are "orientation" and "magnitude." The orientation property of a vector at any given point in a potential field exclusively indicates direction and alignment of the robot's motion and vector's magnitude denotes the extent to which the robot is guided along that orientation. As a case in point, Figure 66 illustrated a potential field around an obstacle, in which the position of the robot is specified. Vectors' direction is in such a way as to culminates in making the robot leaves the obstacle behind. On top of that, the magnitude of vectors in the vicinity of barrier is enlarged, and at further distances away from the obstacle; this magnitude is decreased.

Calculating Potential Field Vectors at Target and Reference Points

Procurement of potential field vectors at reference and target points is a fairly uncomplicated process. Potential field vector in each point of the determined area can be specified as follows:

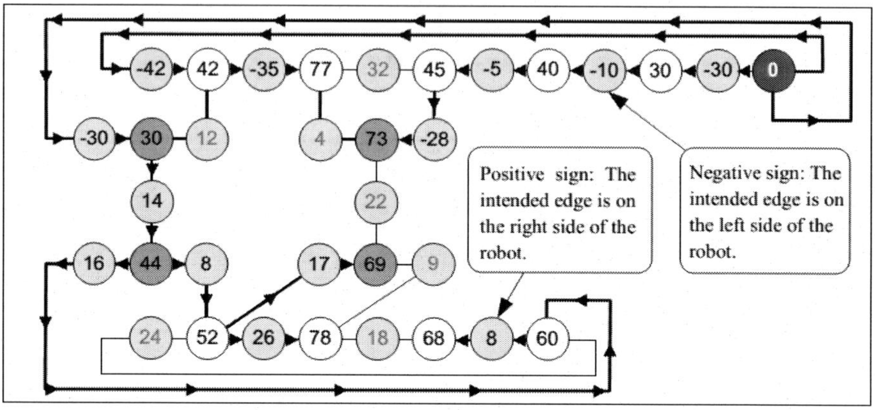

Figure 64. Guidance graph of choosing path.

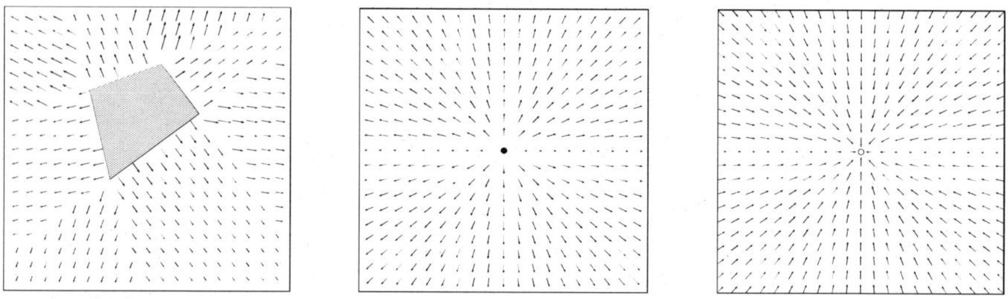

Repletion from obstacle | Repletion from reference point | Attraction towards the target

Figure 65. Three principles of changes in robot path in the potential field.

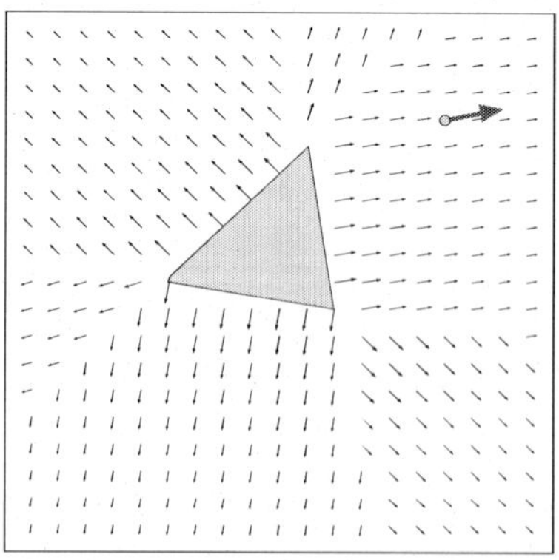

Figure 66. Vector orientation and magnitude in potential field represent the extent of its orientation and importance in guidance.

$$\vec{v}_g = v_{g,mag}(v_{x,g}\boldsymbol{i} + v_{y,g}\boldsymbol{j}) \tag{12}$$

in which $v_{g,mag}$ refers to the magnitude of potential field vector and $v_{x,g}$ and $v_{y,g}$ stand for this vector components in the target position. By knowing the direction of the resultant vector which must be pointed from the coordinates (x, y) towards $(v_{x,g}, v_{y,g})$, these components can be procured as follows (Figure 67a):

$$\begin{cases} v_{x,g} = g_x - x \\ v_{y,g} = g_y - y \end{cases} \tag{13}$$

The magnitude of the target vector $v_{g,mag}$ at any position is quantified as follows:

$$v_{g,mag} = \frac{\alpha_{goal}}{\sqrt{(g_x-x)^2+(g_y-y)^2}} \tag{14}$$

where α_{goal} is defined as a constant amplification factor of target vectors, the value of which can be determined in proportion to the complexity of the environment in addition to the number of obstacles.

Procurement of potential field vector at the reference point is also akin to the calculation made for the target point, except that in this case, the direction of the vectors is directed from the reference point (r_x, r_y) towards the point (x, y), as shown in Figure 67b. Depending on the case, amplification factor at the reference point of vectors could even be calculated using the resultant potential field irrespective of the target point in order for the robot to be navigated in a proper attitude towards the target position.

Calculating Potential Field Vectors of Barriers

The quantification of potential vectors for the obstacles is a slightly more elaborate procedure. Independent vectors need to be established for every edge of an obstacle separately, and all of them should be combined in the next step. Each edge of this barrier splits the plane into two regions, one of which encircles the obstacle, and no part of the barrier is introduced to another side (Figure 68). Potential vectors for each edge are solely formed in the side that lacks an obstacle. The position of a given point germane to every edge of the barrier can be attested with a series of clockwise vectors, manifested in Figure 68.

Reasoning from this fact, any point of the plane inside which a counter-clockwise composite edge along with the vector is formed, will be influenced by that edge. Therefore, a non-zero vector is determined. The vector will be zero for the remaining points of the plane in relation to this edge. That is to say, direction of rotation at the corresponding points (x, y), (d_x, d_y) , and (s_x, s_y) in Figure 68 is counterclockwise. A vector for this direction should be specifically delineated. Type of rotation can be procured applying the right-hand rule and employment of determinant operator as follows:

$$\begin{vmatrix} s_x & s_y & 1 \\ d_x & d_y & 1 \\ x & y & 1 \end{vmatrix} = s_x d_y + s_y x + d_x y - d_y x - s_y d_x - s_x y \tag{15}$$

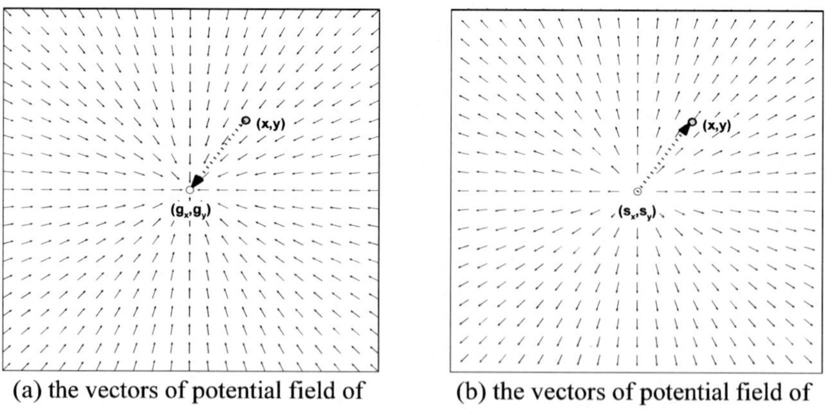

(a) the vectors of potential field of target point

(b) the vectors of potential field of reference point

Figure 67. The position of the field points comparative to the reference point and the target.

according to which, based on the value of resulting determinant if:

- This value is negative; it indicates counterclockwise;
- The value is positive; it implies clockwise;
- The value is zero; it stresses alignment of those two vectors.

It has become evident that a point on the plane should be dominated by an edge; its vector can be determined by means of the following formula:

$$\vec{\boldsymbol{v}}_{obst} = v_{obst,mag}(v_{x,obst}\boldsymbol{i} + v_{y,obst}\boldsymbol{j}) \quad (16)$$

in which, $v_{obst,mag}$ indicates the vector's magnitude, and $v_{x,obst}$ and $v_{y,obst}$ form the unit vector components of the potential field for each edge that is computed having regard that the resultant vector must be perpendicular to and away from the edge, as following equation (Figure 68):

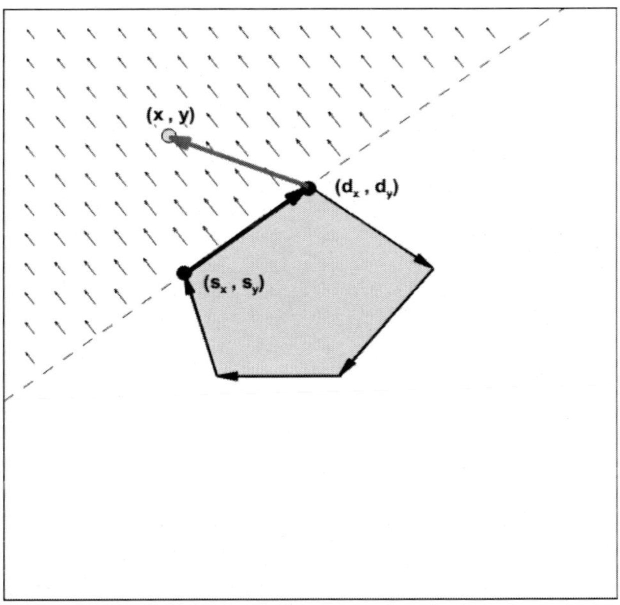

Figure 68. Vector representation of a barrier with a collection of clockwise vectors.

$$\begin{cases} v_{x,obst} = \frac{s_y - d_y}{L_{sd}} \\ v_{y,obst} = \frac{d_x - s_x}{L_{sd}} \end{cases}, L_{sd} = \sqrt{(d_x - s_x)^2 + (s_y - d_y)^2} \quad (17)$$

In a similar way, $v_{obst,mag}$ alludes to the magnitude of potential field vector and can be procured as follow:

$$v_{obst,mag} = \frac{\alpha_{obst}}{1 + \frac{d}{\varepsilon \alpha_{obst}}} \quad (18)$$

in this relation, d means the minimum distance starting from the point (x, y) directed along the edge, α_{obst} demonstrates the obstacle's amplification factor, and ε determines the growth factor of vector's magnitude comparative to the rate of increasing distance from the edge. The amount of d is contingent on whether the point (x, y) is included within the range of values perpendicular to the edge or located beyond this range. Consequently, d is computed based on equations (19) and (20) respectively:

$$d = \frac{|ax + by + c|}{\sqrt{a^2 + b^2}}, \quad \begin{cases} a = d_y - s_y \\ b = s_x - d_x \\ c = bs_y - as_x \end{cases} \quad (19)$$

$$d = \min(\sqrt{(x - s_x)^2 + (y - s_y)^2}, \sqrt{(x - d_x)^2 + (y - d_y)^2}) \quad (20)$$

Parameters α_{obst} and ε can be evaluated with respect to the complexity of the environment and the number of obstacles. Various potential fields are constructed with modification of these parameters whose impacts on the potential field of an edge are reflected in Figure 69. As has been shown, the more α_{obst} escalates under constant ε, the larger the magnitude of vectors and as ε magnifies under constant α_{obst}, the amounts of vectors can be lowered with increased distance from the edge.

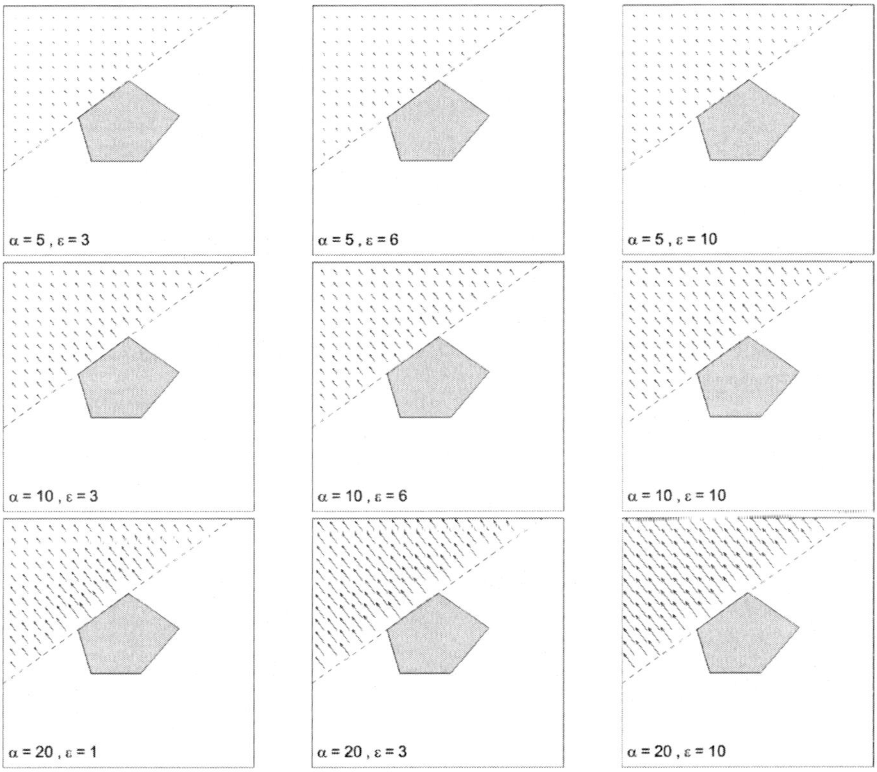

Figure 69. The impact of the parameters; coefficient of amplification (α) and ε, the growth factor, on the potential field of a barrier.

Combining Potential Vectors

In the previous section, the procedure of the potential vectors' formation for the different components of an environment was explicated. These independent components, which encompass from the reference and the target points to the barriers' edges, can be combined applying the vector sum to derive the final potential field (Figure 70). To facilitate the union of these two vectors, they are decomposed into their components along y and x directions, so their combination can be obtained.

An example of combining vectors of potential fields that is pertaining to the edges of a barrier is illustrated in Figure 71 whereby the potential field of edges is specified independently. Furthermore, the combination of the potential field of the obstacle's edges is depicted in Figure 72. As can be observed, the situation or status of the vector at every point in the field relative to the barrier is particularized well in combinational potential field method, and if the robot locates in the proximity of that barrier, it is able to detect its direction motion far from the barrier.

Another exemplification of combinational vectors is a complex barrier having manifold edges, and convex and concave corners, the resultant potential field of which is formed from combining the potential field of all edges, and is displayed in Figure 73. In this potential field, the more the distance from barriers, the substantially less the vectors' magnitude, this means that a small amount of number ε determines the vectors of this field.

Path Planning Based on Combinational Potential Field Vectors

The robot's route from starting point to the target point is procured by the combination of potential fields from different components including barriers, the beginning, and target points. In a similar case, Figure 74 demonstrates an environment possessing five components; three barriers, and the reference and target points. As indicated, the combinational potential fields of these components have led the robot to choose the proper direction for its path towards the target with the simplified mathematical computations.

In some conditions, there appears a repulsion force in the field at the starting point, which represents a reasonably tough challenge to arrive at a solution. An illustration of such kind of field is provided in Figure 75. As can be seen, the vectors near the reference point take a considerably small magnitude and seriously impedes the robot's progress towards the target point. In certain instances, the amplification factor of repulsion force exerted from barriers could be to such an extent that the combination of vectors around the reference point converges towards this point in lieu of proceeding in the opposite direction far from it. In cases like this, the robot guidance towards the target hinges on the strength of the generated field.

Figure 76 exemplifies a field with average attraction strength and Figure 77 exhibits a field with high attraction strength that can be formed by enlarging the amplification factor at both target and reference points. However, receiving an excessive boost in field strength can culminate in a collision between the robot and obstacles. An illustration is provided in Figure 77, according to which the resultant vector that is derived from the combination of edges of barriers magnetizes the robot to the barrier in place of pushing back the robot to be away from the barrier due to the orientation vectors of some edges; therefore, the robot would be on a collision course with the barrier.

Local Minimum

In some cases, the vectors' magnitude in the proximity of the reference point even when combining with other components of the field is considerably insignificant so that determining the path in practical terms is unattainable. In Figure 78, this state of the potential filed, which is called local minimum, is demonstrated. Two methods to evade being captured in a local minimum:

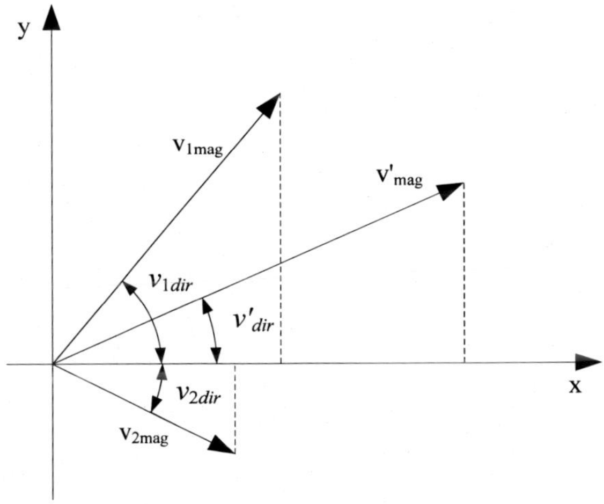

Figure 70. Vector sum of two vectors in the potential field.

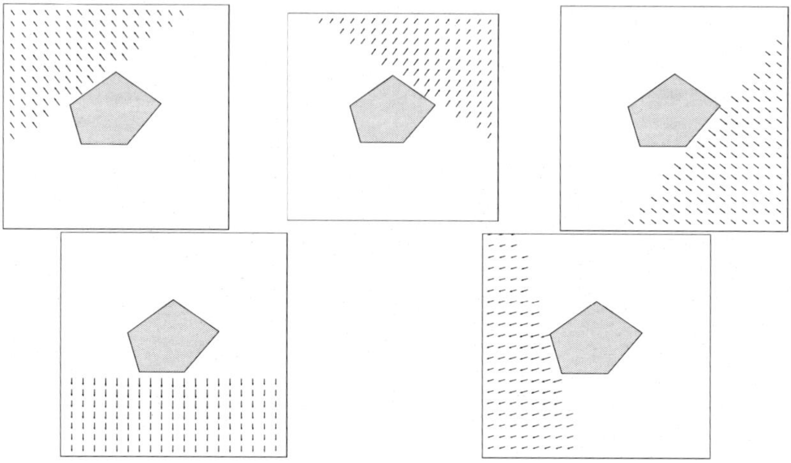

Figure 71. Independent demonstration of potential fields adjacent to the edges of an obstacle.

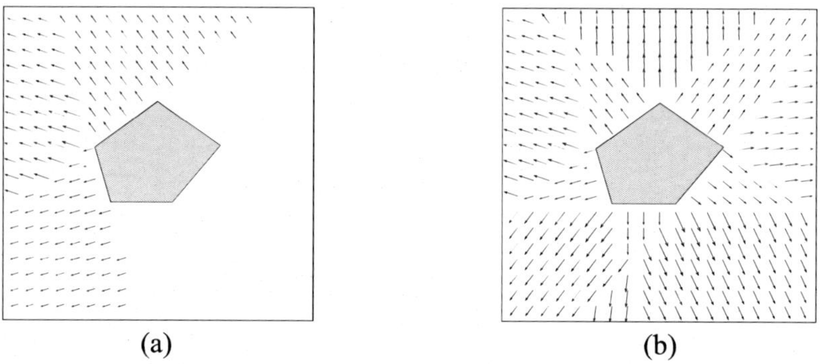

Figure 72. Combination of vector fields near the edges of an obstacle; a) two edges b) all edges.

- The intensity of the potential field at reference point should be strengthened
- Noise needs to be introduced to the environment

The first approach deals with the addition of some intensity value to the reference point, from which the increased repulsive force is exerted to the robot and leads it to not being static (Figure 79). In the second method, the introduction of some to the environment can be served as a vector whose size is fixed, and its direction is random (Figure 80).

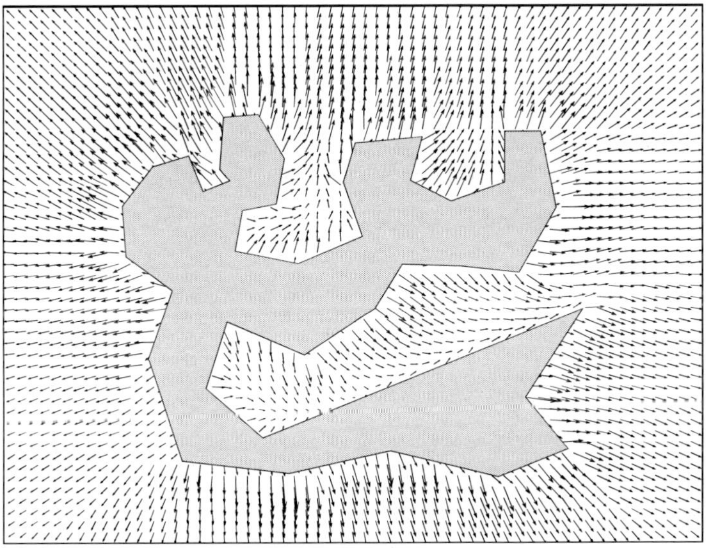

Figure 73. Applying left turn method for unjust obstacles.

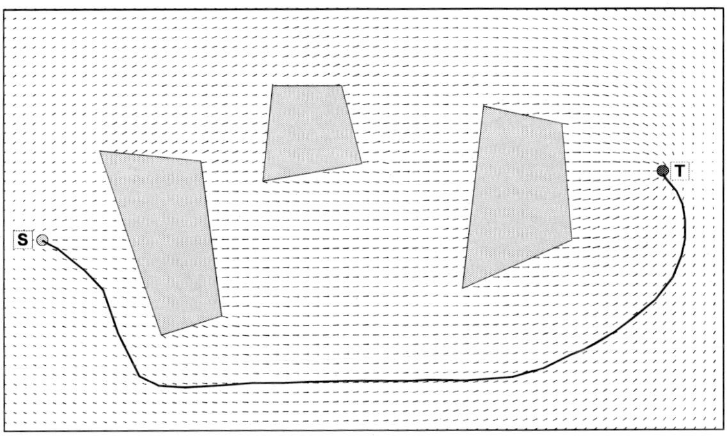

Figure 74. Combination of potential fields in an environment including a few obstacles, to determine the direction of the robot.

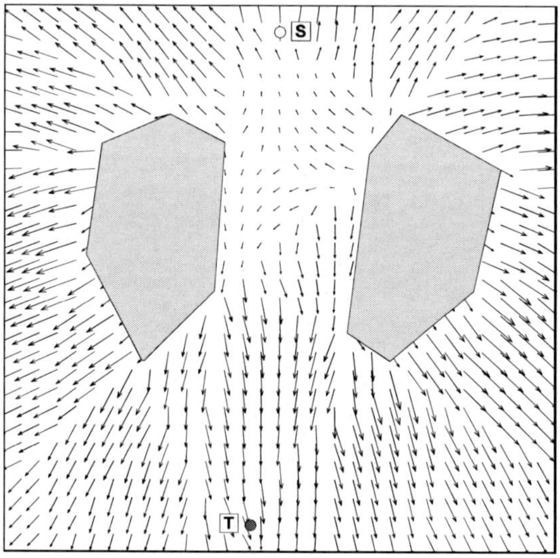

Figure 75. A potential field with a weak target attraction.

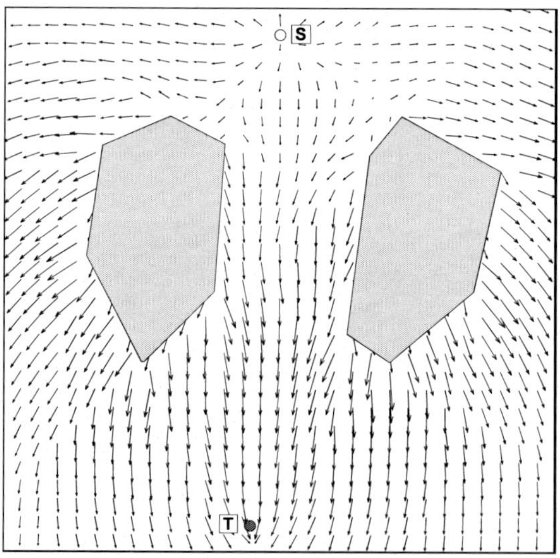

Figure 76. A potential field with a mediocre target attraction.

Enhancement of the strength of the reference field on many occasions does not result in the creation of a path which guides the robot toward the target, and in some cases makes the situation worse and causes the robot being trapped into the local minima. On the other hand, noise adding in general (of course not always) bring about both surmounting the local minimum and detecting a path that stirs the robot towards the target. Figure 81 illustrates an example in which the noise field created a tremendous impact on opening up a path towards the target.

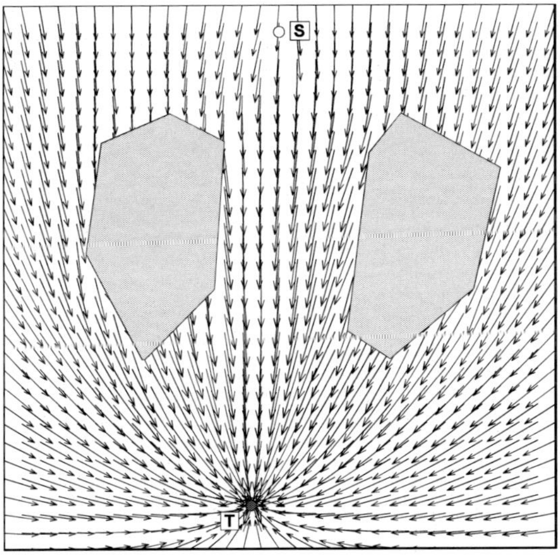

Figure 77. A potential field with a strong target attraction.

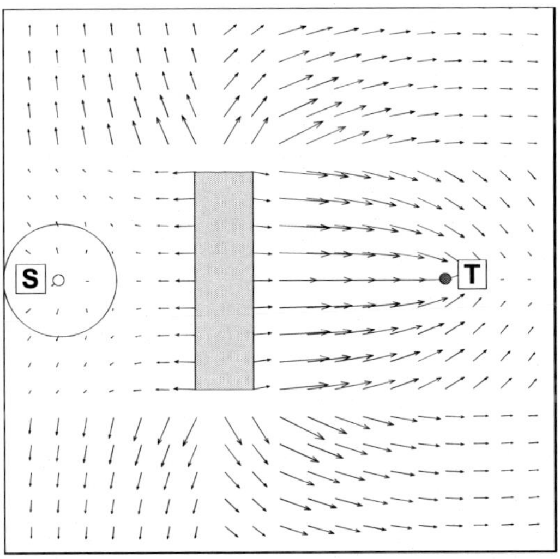

Figure 78. Existence of a local minimum around the reference point of a potential field.

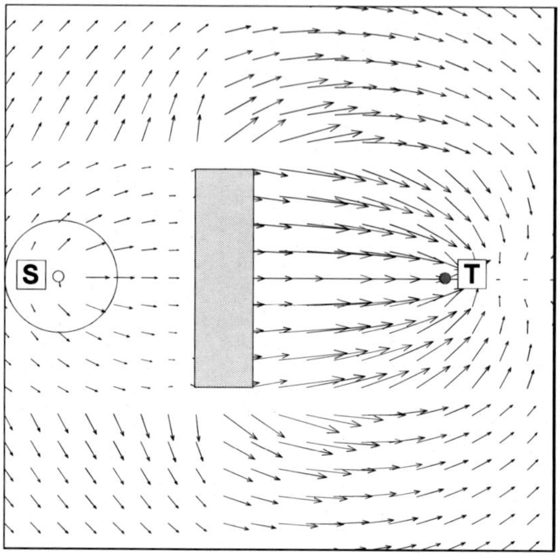

Figure 79. Prevention of getting caught in a local minimum by creating repulsive force from the reference point.

Figure 80. Creating noise (vectors of the same size with random orientation) to prevent being trapped in local minimum of a potential field.

It is worth noting that the motion path is permanently alternating in harmony with the random noise vectors during the course of the robot's searching for the target point. Figure 82 perfectly alludes to this status. Besides, the inordinate amount of noise vectors gives rise to their ineffectiveness, ensuring that the robot is not able to detect its path for attaining the target. This type of case is demonstrated in Figure 83.

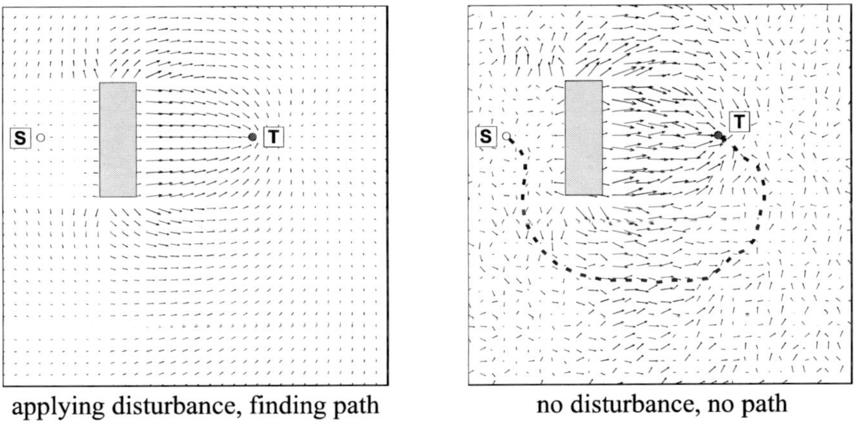

Figure 81. The impact of creating noise on finding the best choice of reaching the target.

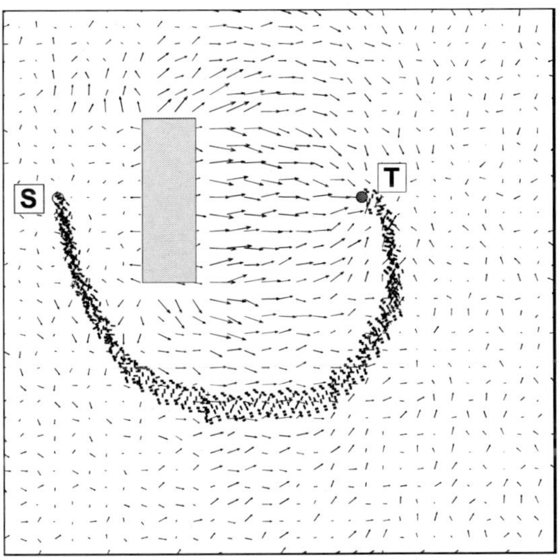

Figure 82. Continuous change of the path by changing the noise vector during the robot's movement towards the target.

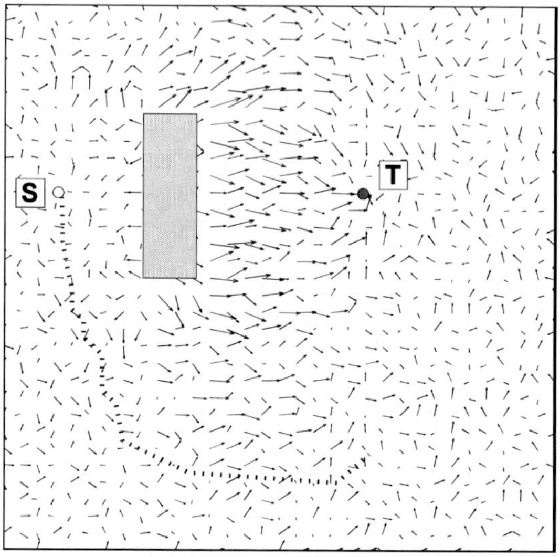

Figure 83. Failure of the robot to find the path due to excessive noise.

Chapter 4

Path Planning and Robot's Hardware

Introduction

In this chapter, path planning is going to be dealt appertaining to the robot's sensors and computing hardware. In unknown environments, the map of which does not initially exist, the sensors are responsible for the accumulation of the required information. Path planning dependent upon robot hardware configuration can be as the following:

Path planning based on mapping functionality:

- With the capability to map the navigated path without memory (without a map)
- With the ability to plan the environment

Path planning based on sensor data:

- Distance sensors: ultrasound and laser sensors
- Attendance sensors: infrared and contact sensors
- Vision sensors: single-vision, bi-vision, and multiple vision sensors
- Combinational sensors (data fusion)

The first section of this chapter takes up the introduction of path planning in terms of the mapping capability. In this section, three mapping approaches such as polygon, gridding, and hierarchical mappings will be expounded, and the pros and cons for each are going to be inspected. The second part delves into path planning based upon visual sensors in particular with regard to the three methods namely “distance-finding”, “attendance-finding”, and “vision-based”, each of which possesses its own special features. Vision sensors turn out to be best yet due to their information that is stored at their disposal for path planning matters. This section will provide further clarifications on the nature of light, and be followed by the description of three successive stages, accurately recording, storage and recognition of light employing the visual sensors. In the fourth section, the motion of two-dimensional robots encountering with two-dimensional obstacles located in their surrounding environments would be explicated [59]. Be more precise, these types of robots may have two or three degrees of freedom (i.e., translation along directions x and y and/or rotation towards θ). Further in this section, the way to model robots in the mentioned environments will be delineated. The terse explanation of two well-established and standard algorithms whose name are configuration space (or empty space) and potential field are outlined, according to which, a combinational strategy enjoying high speed and reliability is presented and after all its implementation results in miscellaneous situations are reflected.

PATH PLANNING BASED ON MAPPING CAPABILITIES

The environment in which the robot is positioned may undergo changes in the course of time. That being so, a map is an appropriate tool to draw the environment at specific times. On this account, the robot should possess multiple abilities, enumerating planning guidance strategies, prevention of colliding with obstacles during its motion, identification of modifications in the environment, recognition of accessible and inaccessible zones, and the capability of pinpointing its current position on the environment.

Maps are an approximate representation of the environment indeed. This approximation seems to be imprecise or even erroneous in many cases. An example of such a faulty map may be perceived in Figure 84. Consequently, the robots are required to depend on their sensors, together with the map to keep away from collisions.

The maps can be presumed to be accurate, legible and constant (its values are immutable) over time (time-invariant), and this section then comes with instructions to plan paths working under this assumption. The next section goes into the way through which a map is extracted from an unknown environment. In order to enrich the understanding of the discussed points, diverse types of maps are brought up as:

Figure 84. An example of a map that provides an uncertain estimate of the environment.

- Geographical maps: The relation between the obstacles is typically recorded using this kind of maps. In some cases, they record voids within the environment instead.
- Obstacles maps: the position of obstacles and unreachable locations is recorded applying these maps.
- Outdoor maps: one of the potential applications of these types of maps is the determination of the sound zones for the robot displacement in an environment,
- Path maps: the onus is on path maps to record the routs in which the robot is able to proceed with safety. Besides, these maps can be used for industrial application as well as performing the guidance of the robot in known paths.
- Hierarchical maps: These kinds of maps which are generally signified as locational maps, record relations between obstacles or cells in order to represent the connections between various parts of an environment.

In this book, the study of two-dimensional maps (such as Figure 85) will suffice.

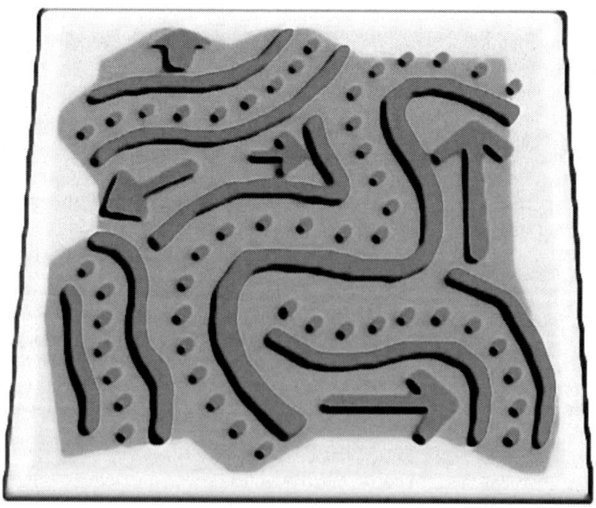

Figure 85. An example of a two-dimensional map of the environment.

Mapping Obstacle Maps Using Polygonal and Gridding Methods

Preparation of obstacle maps is made in two methods: In the first method, polygons are formed from straight lines to detect obstacles boundaries. On the other side, in the second method, the environment settings are grid in the shape of a two-dimensional image whose grid cells are used in order to identify the obstacles. A case in point which is adopted these two methods for the same environment is exemplified in Figure 86.

The fundamental distinction between these two methods is germane to the memory consumption for storing the image, algorithmic complexity, and run-time of the algorithm. In vast areas with an insignificant number of obstacles, the first method known as vector storing requires tiny memory space to record an image. The second method, nevertheless, memory space is taken up irrespective of whether the cells are void or full; under this account, storage of the image demands a pretty colossal storage space. Despite this fact, there is no question that the second method (environment gridding) seems to be more preferable in environments with many but small obstacles. From this perspective, the difference between these two methods is contingent upon the environment in which they are employed (with low or many numbers of obstacles), as can be observed in Figures 87, and 88. following after.

At the same time, the memory requirements for storing the image by gridding method, which is named raster maps, is closely related to the size of the network. An example being, the requisite memory to record a grid with the size of $M \times N$ is at the scale of $O(MN)$. In this manner, vector maps in each grid with small size $(m, n \ll M, N)$ demonstrate better performance. By Supposing m obstacles, each of which has n vertexes, the space necessary to save the image via vector method is procured based on the following equation:

$$storage = (m \times n) \times 2 + 2 \times (m \times n) = O(mn) \tag{21}$$

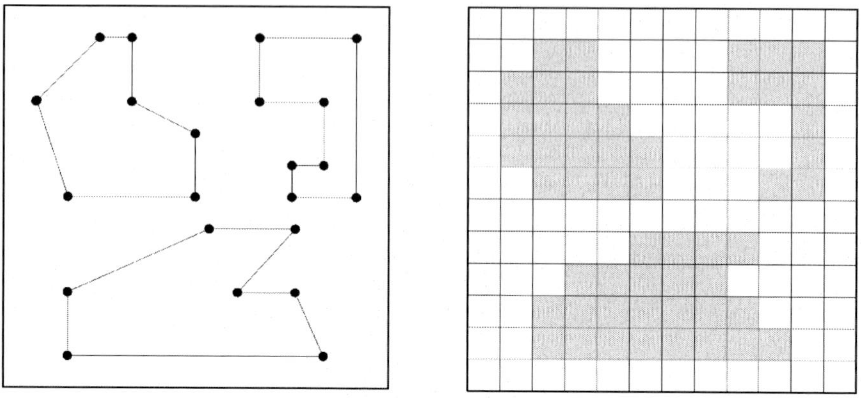

a) Mapping based on the polygonal method

b) Mapping based on the gridding method

Figure 86. An example of a map of the environment and obstacles apropos of the polygonal and networking methods.

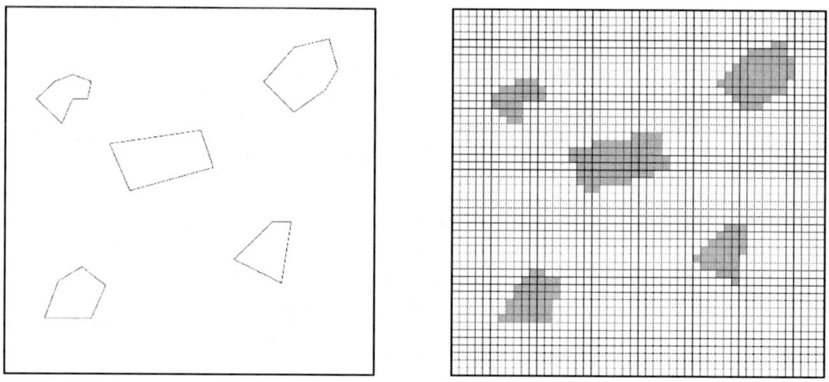

a) Polygonal method

b) Gridding method

Figure 87. Polygon and networking maps from an environment with a low number of obstacles (less required memory for the polygon method).

The resolution of an image in gridding method hinges upon the accuracy. By evaluation of Figure 89, it might be understood that the difference arises out of resolution. It is of crucial importance for storing the maps to reduce the resolution of images during the reduction of the memory, in turn, leads to precision decline. This loss carries the implication of problem-solving procedure. For example, in an environment, resolutions of which are low, medium, and high, as can be seen in Figure 89, determination of the path towards the target might be far-fetched, single-path, or multi-path, respectively. Despite, in the vector maps, the problem-solving process is not contingent on the memory resolution, but the numerical method with a considerable extent.

One of the deployment destinations of implementing grid-based maps seems to be the safety of the robot. The degree to which occupation of each grid cell belongs, foreshadows the presence of obstacles at that particular area of the map. Figure 90 demonstrates three methods depending on the context of obstacle presence. As shown in Figure 90a, in this case, a cell is considered to be filled if no less than 50% of that cell is being occupied by the obstacle, on the contrary as evident in Figure 90b, it is sufficient for filling the cell to be occupied only 1% of its space. In Figure 90c amount of cell space occupied by obstacle is stored as a piece of information in that specified cell. In practical terms, the vast majority of robots exploits the sort of maps displayed in Figure 90c. This map provides the opportunity to use fuzzy algorithms in order for the position of obstacles to be determined. Therefore, the level of occupation for each cell indicates the possibility of an obstacle presence into that cell.

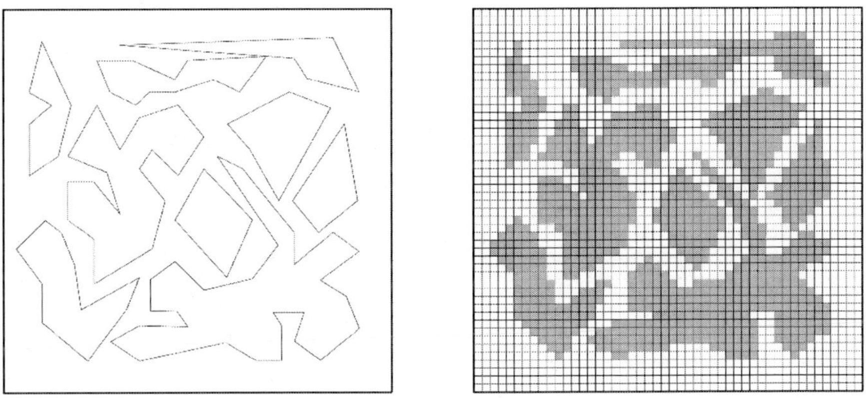

a) Polygonal method b) Gridding method

Figure 88. Polygon maps and networking from an environment with high number of obstacles (less memory for networking).

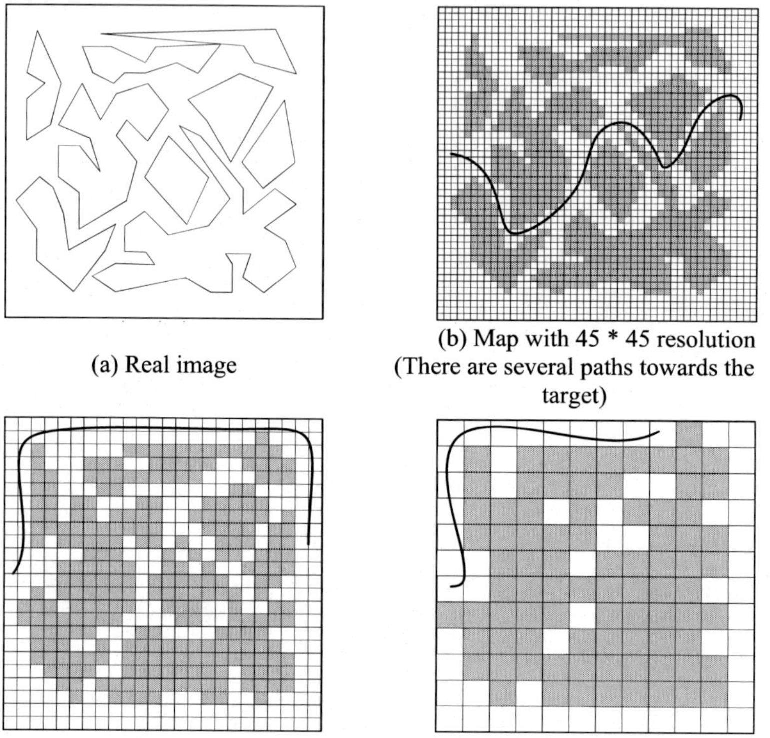

(a) Real image

(b) Map with 45 * 45 resolution (There are several paths towards the target)

(c) Map with 24 * 24 resolution (There are two paths towards the target)

(d) Map with 24 * 24 resolution (There are two paths towards the target)

Figure 89. Changing the number of paths to the target in maps with different resolutions.

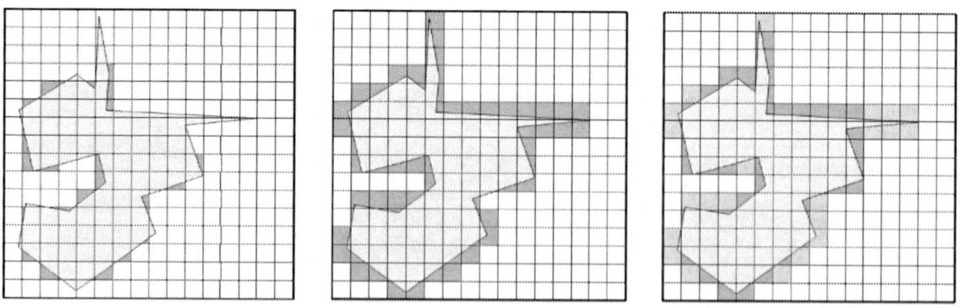

Gridding is based on occupying more than 50% of cell space

Gridding is based on occupying at least 1% of cell space

Gridding with considering the amount of space occupied by each cell

Figure 90. Determining the presence of an obstacle in a cell from the network using its occupancy rate with the obstacle.

Hierarchical Maps

Another type of maps, which are titled as hierarchical, hoards relations between obstacles or cells. Two types of which called "topological maps" and "feature maps" tend to be instituted in this section.

Topological Maps

Topological maps provide the representation for the relationship between the separate division of an environment. Examples include, an obstacle might be found in an environment, one or more hole of which is spotted in its betwixt (Figure 91). It goes without saying that the robot is needed to desist from approaching these apertures or cavities. In vector maps, if there exists an opening between the obstacles, the robot is permitted to proceed towards it.

Topological maps in which these holes are demonstrated as empty cells make use of a structure named "square tree" to diminish the requisite memory. According to this structure depicted in Figure 92, horizontal and vertical lines intersect the environment into four equal sub-zones, and gridding is solely accomplished in areas where obstacles are present. The resultant consequence of this subdivision is a tree-like structure akin to Figure 93, in which contiguous cells should be read from left to right.

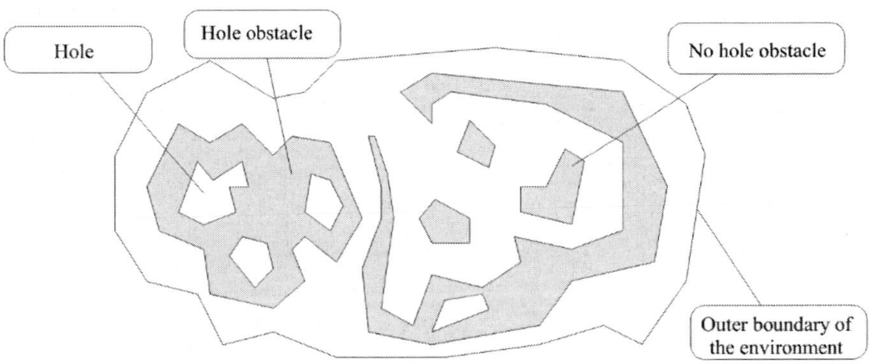

Figure 91. Obstacles with or without holes in topological maps.

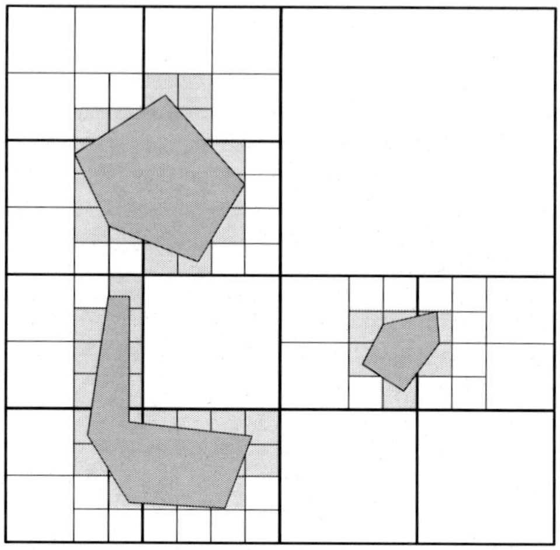

Figure 92. Demonstration of the rectangular tree's performance in topological maps to divide the environment.

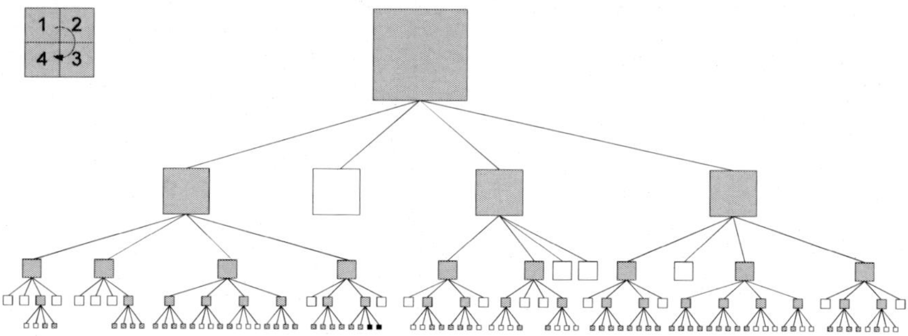

Figure 93. The tree-like structure of the environment cells in the topological maps.

In environments with low-density obstacles (non-cluttered environments), the square tree method is able to reduce the memory requirements of accumulated images. By virtue of regarding cells with varying sizes in this method, deployment of this method demands specific programming configurations in the path planning algorithm.

Feature Maps

Feature Maps as being a form of hierarchical maps is capable of storing environment properties viz free spaces or boundaries. Figure 94 illustrates a representative case of maps that were produced through following boundaries of obstacle and storing the angles of corners and the lengths of edges without having to collect information about exact coordinates. The strategy with which preparing his types of maps is provided was fully characterized in Chapter 2.

The accuracy and precision of drawn maps hinge on the parameters among which the minimum rotation angle of the robot stands out. This characteristic of the robot comes to light at the corners of the map. For instance, as shown in Figure 95, the minimum rotation angles of 15, 45 and 60 degrees from which exploitation, preparation of a feature map is commonly made. The wider the rotation angle, the lower the level of accuracy of storing the corners and edges, to the extent that it gives rise to obstacle collision for the robot.

On the other hand, various stepping approaches across the obstacles may even contribute to the formation of separate maps. Figure 96 gives a demonstration of two different environments based on which maps were prepared by traversing the environments once in a clockwise and yet again in a counter-clockwise direction. As is evident, recording corners and edges differ sharply in the two stepping methods.

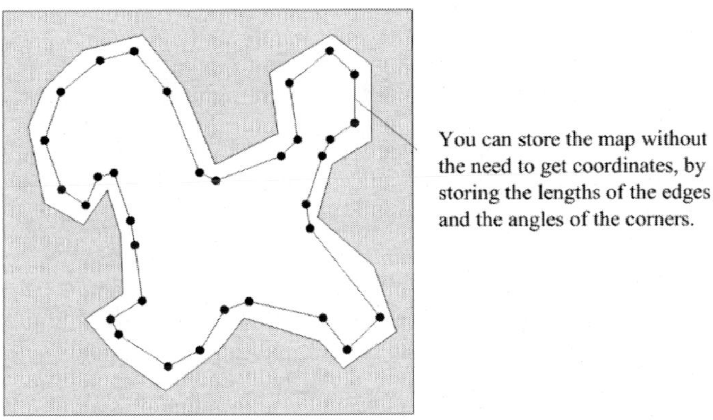

Figure 94. Recording corners and edges via tracking obstacles.

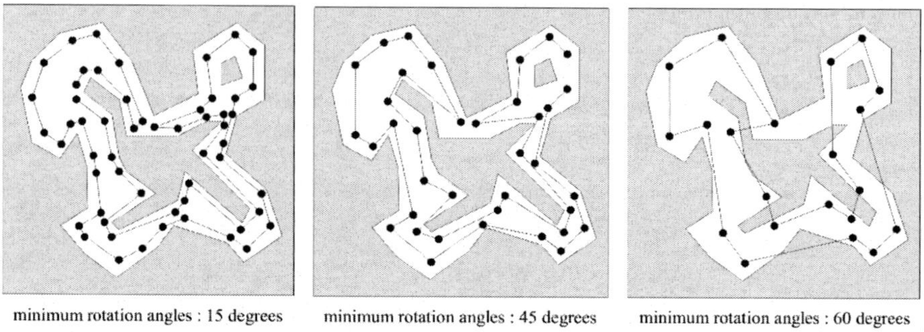

Figure 95. The effect of the minimum angle of rotation parameter on the accuracy of feature maps.

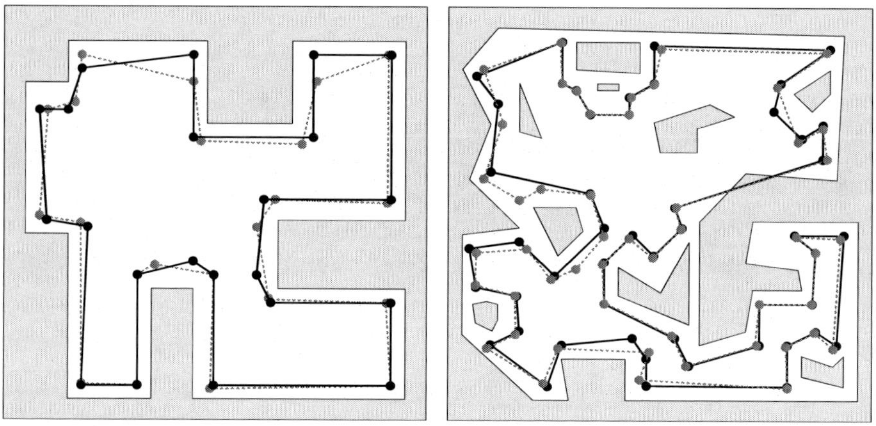

Figure 96. Comparison in the map taken from clockwise and counterclockwise shooting in the two environments mapping.

The robots perplexed in the time of its navigation towards edges in symmetrical environments. In Figure 97 as a case in point, there has been a number of obstacles whose edges are symmetrical. When passing these obstacles, the robot gets confused (marked with a thick line).

On the flip side, unless an external source or an extensive reference coordinate system is provided at the robot's disposal, it appears incapable of identifying the rotation direction of the edges in global coordinates. As an exemplification, three environments of (a), (b) and (c) portrayed in Figure 98 happen to resemble one another with respect to the length of subsequent edges for the robot; inasmuch as the sole contrast between these environments relates to a rotation when a general coordinate is lacking, it becomes unfeasible for the robot to draw a distinction between these apparently similar environments. The position of the interior obstacles in the environment (d) is not, nevertheless, a form of rotation occurred for other three environments. For this reason, the robot does not get into difficulty to distinguish this environment in contrast with others.

In addition, relative rotation and comparative location can be detected with no coordinating information and by entirely banking on the footprint of the robot through just a few broad rotation parameters, for example, owing to the resemblance of the obstacles and inaccessibility to coordinate information, maps of Figure 99 in the environment becomes not inconspicuous for the robot. Under those circumstances, according to Figure 100, in addition to the feasibility of considering the effect of rotation, the obstacles' distances in the environment can be surmised by erecting several bridges or shortcuts between the barriers.

In a nutshell, it is worth mentioning that drawing detailed maps whereabouts the robot follows a quick route, would be justified, having said that at great distances, due to the necessity for high memory and also ever-increasing processing load to discover the optimal path, these maps cannot be put into practical implementations. For medium distance which runs between the gamut of the mentioned extremes, mapping the traversed passage seems to be sensible.

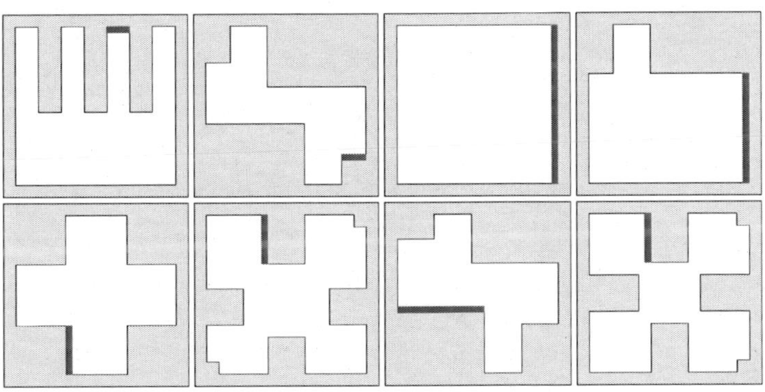

Figure 97. Some examples of symmetric environments.

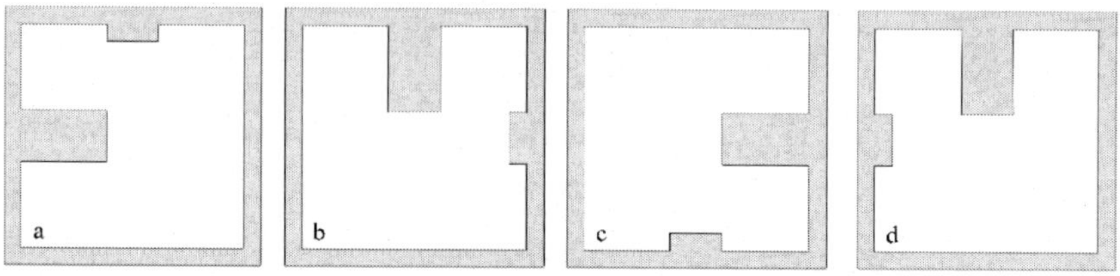

Figure 98. The rotated environments with (a, b, c) analogous lengths and (d) different lengths.

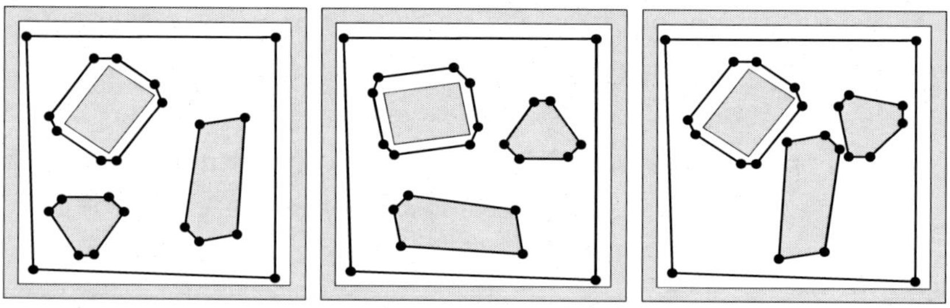

Figure 99. The demonstration of the impossibility of determining the wide rotation, relative rotation, and relative location.

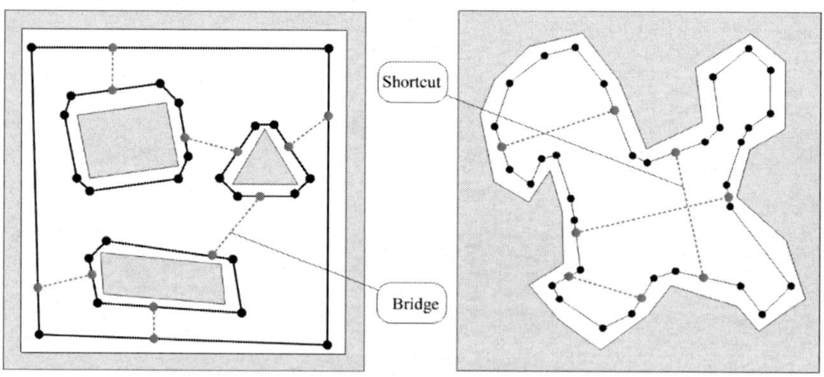

Figure 100. The illustration of shortcut and bridge in the two different maps.

ROBOT PATH PLANNING USING VISION SENSORS

As has already been pointed out in the introduction of this chapter, routing corresponding to sensors comprise of three methods in particular "distance", "attendance" and "vision", each of which has its own distinct attributes. Attendance sensor predominantly elongates the path and provides the robot with the least information compared to the other methods. Accordingly, methods that are dependent merely on sensors are outdated. However, the distance sensors provide the robot with extensive data for routing; the derived data from these sensors is unhelpful for robots to be aware of the danger zones. What is more, the operation of active distance sensors in wartime because of becoming perceptible to the enemy consider to be as an incautious approach. Visual sensors (cameras) has become overwhelmingly dominant sensors within the routing framework for the information they deliver. This topic will be presented in detail in Chapter 5.

ROBOT PATH PLANNING USING RASTERIZING

Movement of two-dimensional robots towards the target in environments with both big and small obstacles considers as one of the dramatic issues in path planning problems. In such problematical matters, the robot is normally modeled as a point which contributes to transcending the barriers of the two-dimensional robot. In order to grapple with this limitation, a number of adjustments and rectifications are also made for modeling the environment, and inter alia, obstacles, to avoid the robot's potential colliding with obstacles. Two-dimensional robots can have either two or three degrees of freedom. One of the classical versions in the field of path planning is referred to the problem that has been solved with different algorithms, among which two configurations, namely space, and potential field methods has gained increasing acceptance in contrast to others. In this section, each of these two above mentioned algorithms is described in full detail, the superiorities and limitations of which will be then mentioned and at last the adjusted path design algorithm on the basis of configuration space which being of higher speed and flexibility is put forth. This algorithm can be run in a different environment in which robots and obstacles are allowed to be of any arbitrary type of polygonal geometry, comprising both discrete components and convex angles. According to this algorithm, if there exists a route towards the target position, it will definitely detect such a path. The final section of this chapter provides several examples of the execution of this rectified algorithm in different environments and circumstances in order to demonstrate its operative capabilities.

Configuration Space Method

Definition of Configuration Space

A two-dimensional object with three degrees of freedom moving on a plane surface is supposed. The position of this moving object based on its coordinates along x- and y-axes and orientation towards θ can be achieved. The configuration space for this moving object on a plane surface is the set of all possible situation with three configuration parameters (x, y, θ) where $(x, y, \theta) \in R^2$ (real plane) and $\theta \in S^1$ (Unit circle). Taking this into account, the configuration space of this movable object can be determined with $R^2 \times S^1$ (Figure 101). In robotics literature, configuration space is demonstrated as *C*-space in an abbreviated form.

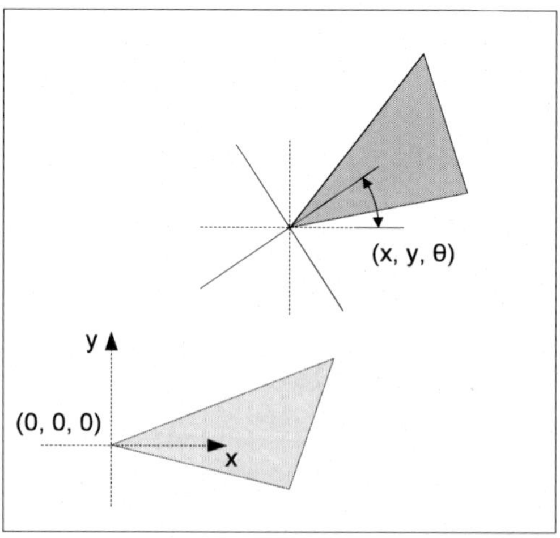

Figure 101. A two-dimensional object in $R^2 \times S^1$ configuration space.

As the next configuration space example, a robot (a dual interface arm) that is connected to a support base is considered at (x, y) coordinate as illustrated in Figure 102. According to this schematic diagram of the robot, the two angles θ_0 and θ_1 which respectively correspond to the base link and free arm angles can completely define the position of the robot. In this system, configuration space is $S^1 \times S^1$, and the point position is specified by $(\theta_0, \theta_1) \in S^1 \times S^2$.

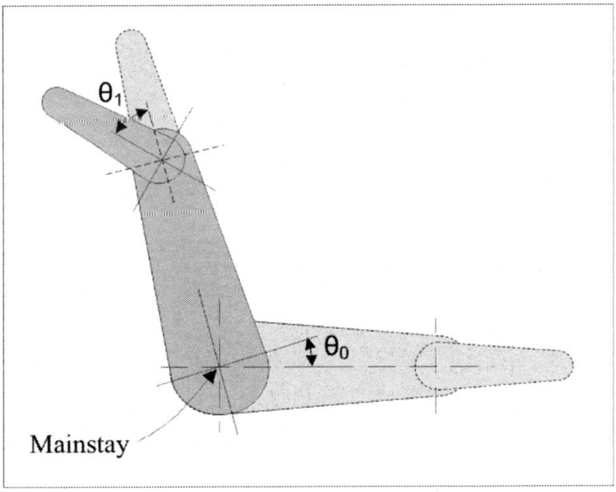

Figure 102. Dual interface arm at (x,y), with the position $(\theta_0,\theta_1) \in S^1 \times S^1$.

Making c-Space Equivalent with Physical Space

The problem of a sophisticated robot's movement in a real physical setting can be considered equivalent to a point's motion problem (this point is usually known as the robot reference point) in the *c*-space. In order for the obstacles' borders in real *c*-space to be tracked and traced in configuration space (*c*-space), obstacles are considered in the broader region of *c*-space; this implicates that enlarging obstacles can offset shrinking the two-dimensional robot to a point. There is no shadow of a doubt that as long as the robot's reference point is situated outside the widened obstacles, the robot itself lies beyond the obstacles in physical space as well.

Lozano-Pérez [60] applied Minkowski summation of a robot and its environment to modify the borders of obstacles in *c*-space. In Minkowski summation, all points of a set are shifted to a new position by another set of points. Minkowski sum is presented as follows:

$$A + B = \{a + b | a \in A, b \in B\} \tag{22}$$

in which a and b denote members of two different sets A and B, respectively. As an illustration, the following two sets are taken into consideration, which refer to the vertices coordinates of two triangles in practical terms:

$$A = \{(1,0), (0,1), (-1,0)\}\ ;\ B = \{(0,0), (1,1), (-1,1)\}$$

Minkowski sum of these two sets will be obtained as follows:

$$A + B = \{(1,0), (2,1), (0,1), (0,1), (1,2), (-1,2), (-1,0), (0,1), (-2,1)\}$$

which is a demonstration of hexagonal form (Figure 103).

Minkowski sum for the sets defined with polygons can be specified just like convolution (set summation) of the obstacle's edges; meaning that the obstacles' polygons together with the negated polygon of the robot can be summed with each other. To give an example, negated polygon of the triangular robot is disclosed in Figure 104a. In order for the robot's environment to be modeled, its convolution with obstacles is used to generate c-space organized as Figure 104b.

Convex hull algorithms can be deployed for quantification of this convolution. The algorithmic time scale for this implementation is $O(n \log n)$ in which n is the number of points. In consideration of foregoing, for a plane without optimum convex polygon, Lozano-Perez put forward an alternative method that the execution time scale of which is $O(n)$. Additional information on the main principles of this algorithm is detailed in [42, 57, 61]. The remainder of this section elucidates the approach for implementing this algorithm.

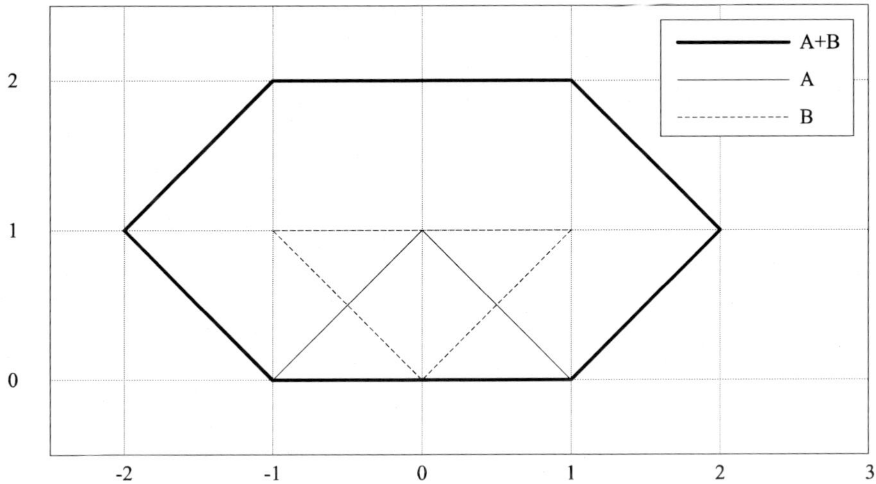

Figure 103. Minkowski summation for the two sets; A and B.

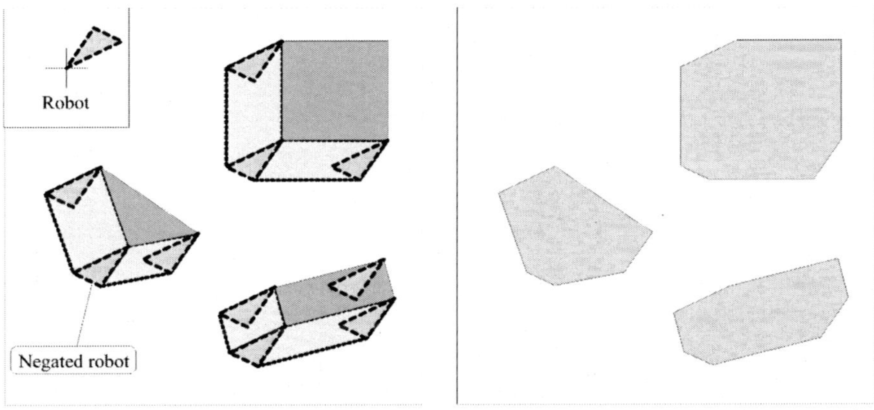

a) Minkowski sum of robot and obstacles b) The polygons derived from Minkowski sum

Figure 104. Minkowski sum of robot and obstacles and the resulting polygons.

It draws attention to what convex set or polygon is meant. This definition can be proposed as a set that direct line connecting any two random points is completely located within which. Otherwise, interest set can be called concave or non-convex set. Two sets of convex and concave are provided as an example in Figure 105.

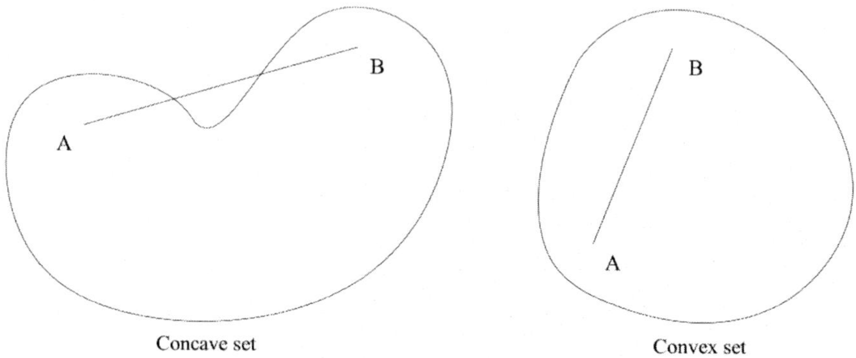

Figure 105. Convex and concave sets.

Rasterizing Methods in c-Space

Other researchers have been motivated to use rasterizing techniques in *c*-space. For example, Lozano-Pérez [60] took advantage of the research carried out by Lozano-Pérez [42], and Donald [62] and encrypted these studies into a bitmap format in *c*-space. However, this algorithm was incompetent for having some problems such as non-local search, heterogeneous, non-parallel-able, and not having any hardware support.

Dorst and Trovato also applied a *c*-space-based rasterized approach to design two-joint arms [63], based on which differential geometric path planning problem with a metric topology has been implemented on the space configuration, and then the shortest correspondent paths with regard to the optimal routes are quantified. With the aim to fulfill this objective, a discrete grid in *c*-space is created at first, after that, cost-wave propagation and gradient-following techniques are exploited so that the optimal routes can be identified. This approach and the third and fourth phases of the algorithm stated in this section closely resemble to each other. The significant discrepancy between this algorithm and the Dorst and Trovato's methodology is that obstacles' creation problem in c-space is not covered in this algorithm, but instead, its emphasis is placed on the hardware aspect, unlike the Dorst and Trovato's work.

Additionally, Donald submitted another motion planning algorithm with reasonable approximation, that getting hold of one route at a minimum (if there exists any) the certain resolution environment is fully guaranteed. Furthermore, all the path(s) proposed by this algorithm is (are) managed to be secure [62]. This algorithm exhibits different characteristics, the most important of which are making use of constraint equations for the representation of obstacles in *c*-space, employing the *c*-space gridding, and after all, implementing the local multi-resolution search approach in c-space. The exceptional merit of this algorithm is that the acquisition of path in a given resolution is verified, but all the same, its software execution, precisely in a non-parallel machine demonstrates a low level of performance. In the remainder of this section, it should be regarded that applying parallel processors can dramatically improve the performance of the algorithm.

Potential Field Methods

Vector Presentation in Potential Field Method

The potential field approach is founded by Khatib and Le Maitre [64]. According to this method, the obstacles are expressed in terms of scalar analytic functions, the value of which should be zero ($f(x,y,z)=0$). Moreover, the local potential field for each obstacle is created through a relation inversely proportional to the square of the distance between two separate obstacles. Furthermore, f_0 as an arbitrary cut value is specified, so that points whose values are lower than f_0 (at a great distance from the obstacle) can be determined (Figure 106a). Wherefore, the mathematical equation of the potential field $P(x,y,z)$ is defined as follows:

$$P(x,y,z)=\begin{cases}\frac{\alpha}{f^2(x,y,z)}, & f(x,y,z)\leq f_0\\ 0, & f(x,y,z)>f_0\end{cases} \tag{23}$$

in which α refers to the amplification factor of the potential field, that is procured regarding whether the corresponding point imposed by the target field is either an obstacle or the robot. Consistence with Newton's laws, a moving particle in the potential field, by no means, hit the obstacle. Khatib in conformity with the law that the summation of gradients is equivalent to the gradient of sum indicated that the total or resultant potential fields of multiple obstacles conclude a unique function. The particle will strike no barrier by virtue of this function. Taking into account that a moving object is not a spot or a point in the real, Khatib prescribed several specified points on the object as a means to employ a linear combination of their potential fields in order that the position and orientation of rigid body can be determined.

The potential field method proved to be successful in dodging transient collisions with obstacles in variable environments, but it bears some limitations in terms of motion design. The principal challenge arises from generating a poor (bad) local minimum more precisely for the convex-shaped robots. In order to evade from the local minimum, either random methods or other approaches should be implemented, some of which are described as follows.

Bitmap Representation in Potential Field Method

Barraquand and Latombe refined the concept of the potential field method and applied bitmaps for the representation of obstacles. As shown in Figure 106b, they constructed a discrete potential field for the robot that will start from the target point [65]. In this technique, the potential field is not a function of distance from one location towards the target, but of the path length that is determined from bypassing obstacles across the route to the target.

Subsequently, akin to Khatib's approach, they integrated the potential field of different points with each other in order to acquire a unique potential field. However, the potential field has no local minimum for a robot considered as a point, the amalgamated potential field for more sophisticated robots might hold some local minimums created in case of distinct points competition. Barraquand and Latombe propounded two strategies for avoiding the local minimum, each of which seeks for a path in *c*-space utilizing potential fields. Filling local minimum as their first method will come with a perfect design in a given resolution. With regard to this, the focused algorithm of interest in this section is closely comparable to their method.

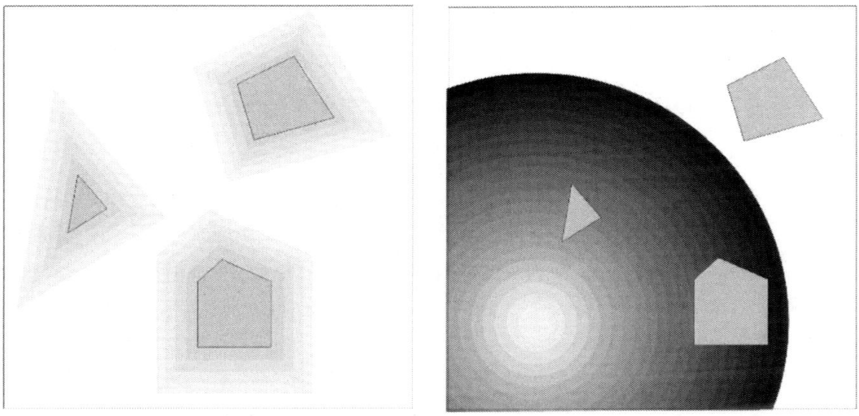

a) Potential field of robot and obstacles   b) Bitmap potential field of target

Figure 106. Potential field resulting from obstacles and the target (The higher the density, the more distinct potential sign).

Combinational Algorithm

Motion design algorithm offered in this section is an ideal combination of all the strength points of the aforementioned algorithms in configuration space and potential field. From an algorithmic perspective, this strategy is formed on the basis of Donald's network-based approach [62] coupled with Lozano-Perez's configuration space method [42] and fabricated of four distinct parts, whose instructions are as follows:

1. Creating *c*-space: in this part, *c*-space is represented through the designation of a voxel (Volumetric pixels), and *c*-space obstacles can be readily evaluated in this array employing standard graphics rasterization hardware.
2. Calculating navigation function: c-space navigation function [66] is computed by a dynamic programming in this part (e. g., expansion of wave-front solutions). The navigation function, which is taking discrete vector values, determines the direction along which the distance towards the target position is contracted,

as a result of having an immediate position of the robot voxel in *c*-space,

3. Determining the optimal path: in this part the shortest path is calculated from each starting point of voxel back to the target (if a satisfactory solution is found). As long as the navigation function computes the direction of motion in every single cell, calculation of the path can be quickly performed in this section, on condition that there exists a route, which the navigation function detects it in a definite time.
4. Drawing motion path: this part relates to the creation of real-time kinematic simulation for the robot movement.

Each of these four parts is going to be fully delineated in the following paragraphs.

Creating Configuration Space Representation

In this stage, the polygons associated with the obstacles of configuration space are quantified via Minkowski sum of obstacles and robots [42], an example of which is illustrated in Figure 104a. Afterward, in order for the polygons of the obstacle's space configuration to be filled, graphics hardware is employed (Figure 104b). Provided that the robot is limited to two degrees of freedom (plane translation), this characterization solely incorporates a single bitmap, let alone creating a space configuration picture. But on the occasion that rotations are allowed, space configuration can be constituted by a set of bitmaps, each of which is made of a sector picture of space configuration into the angular interval as $[\theta_1, \theta_2]$.

c-space representation in one direction can be easily accomplished but from which generating a simultaneous conservative and discrete representation along with a number of angular intervals, is considered to be an arduous task. This approach intends to guarantee that a pixel inside which an obstacle is penetrated, cannot be found to be labeled "free" in a bitmap sector that represents the c-space in an angular interval $[\theta_1, \theta_2]$. With the aim of creating a discrete representation from the *c*-space angular interval of $[\theta_1, \theta_2]$, the *c*-space is divided into *n* parts of equal size, each of which is determined with angular development $\theta = (\theta_2 - \theta_1)/n$. Consequently, all these subintervals are assembled to create the bitmap sector for the angular interval $[\theta_1, \theta_2]$. The pseudo-code for this process is encapsulated as follows:

```
GENERATE C-SPACE ()
For theta = 0 to 2π by 2π/N (N is # of theta slices)
Foreach robotpoly in robot polygon list
  For t1 = theta – dtheta to theta + dtheta
    Rotate robotpoly by t1
    Foreach obstaclepoly in environment
      Generate the minkowski sum of robotpoly and obstaclepoly
      Fill minkowski polygon with obstacle colour.
Read filled polys from frame buffer
Move bitmap into voxel array.
Clear frame buffer.
Return bitmap voxel array.
```

It should be highlighted that bitmap sectors have a conservative manner, meaning that, if a small portion of any given cell is filled with a piece of a penetrated obstacle, all those cells should be labeled "obstacle." Although this task omits a few of potential routes by its very nature, it reinforces the fullness of resolution in the algorithm. Accordingly, the path between the two cells is open if they put together, and they are also labeled "obstacle-free." If that is not the case, motion between them is not permitted, for one or both of them are labeled "obstacle." Put it briefly, the sole movement between free cells is allowed to happen. Upon conclusion, all the resultant routes will be acceptable (collision free).

Calculating Navigation Function

In this procedure, a dynamic path planning algorithm has been deployed for wave-front expansion which emerges from the target configuration and records the positions of the wave-front by lining up in a queue of voxels and forming its characteristics [63]. Each element on the queue is demonstrated by a node within which two attributes are stored, the first of which is the position of a cell from the environment, and second of which refers to the length of its traversed path back to the target point. After each cell occupying an element of the queue, their other adjacent cells that do not exist in the queue before will be added to the remainder of queue.

The first element in the queue, or the element at the front of the queue, is the target cell with a zero-length path. This algorithm excludes this cell from the queue, and its blank position is filled with zero values in the form of pixelated (rasterized) type. Consequently, it sorts all the nearby cells and puts them at the beginning of the queue if the length of their path is 1. Accordingly, the wave-front progresses iteratively until encompassing all the obstacles is met e in *c*-space [67]. It is worth noting that by so doing, every cell that is accessible to the target position should be initialized just for once. The following pseudo-code for implementation of this process, which is shown in Figure 107, is proposed:

```
FILL NAVIGATION FUNCTION (target location)
  Enqueue target location with distance = 0
  While queue is not empty
     Dequeue element F
     Label F´ s location with F´ s distance
     Enqueue all neighbours of F that are not obstacles and that have not yet
     been filled with distance = distance +1
```

The desired duration for running the calculations is proportionate to the free cell numbers in the environment. The more complicated parts of the terrain in which considerable numbers of obstacles exist, the much fewer its free cells, and as a consequence, the more rapidly it will become full.

The process of filling cells in a robot with three degrees of freedom (two translational and one rotational motion) is undertaken in the upwards direction and likewise peripheral direction along θ, just the same as extending along *x*- and *y*- directions.

Path Design

Taking into consideration that the navigation function is designed based upon a Breadth-First-Search (BFS) in the entire *c*-space, the shortest path from the target location to the any given accessible spot in c-space can be detected (a path passing from fewer voxel is considered as the shortest path) [63]. Approaching the target position from any starting point, the robot should proceed to a neighboring cell whose figure is the least. If more than one cell with the lowest figure exists, it is deemed likely that any of them can be chosen; since all of them when considering as origin to start a path have the equivalent numbers of movements to reach the target point. If two options of rotational movement (along θ) and translational motion (along x or y) were available, this algorithm would select the latter option (translational movement) to minimize rotation. The following pseudo-code and Figure 108 implements this procedure:

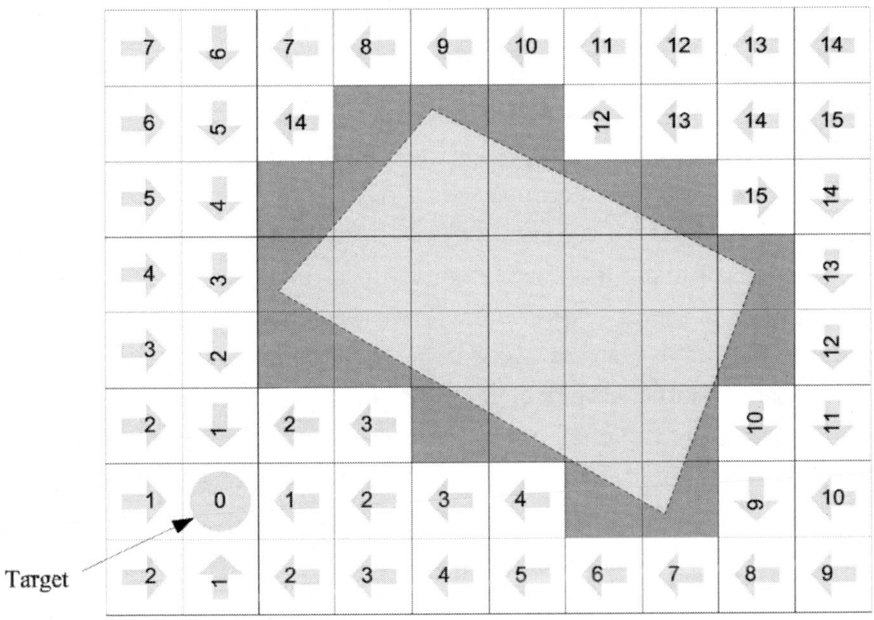

Figure 107. The computation of the navigation function.

```
GENERATE PATH (start location)
   let C = cell corresponding to start location
   let P = NULL
   if C was not reached by fill (label is blank)
      return cannot reach target
   else
      while C is not at target
         add C to path P
         pick lowest numbered neighbour, L
         let C = L
   return P
```

Each cell that is accessible from the target has a neighbor with a lower figure with respect to itself (the cell that has been located in the queue for being labeled). Therefore, there exists no local minimum in the navigation function. The robot is able to track the extensive initial search tree towards the target, without reservation about being stuck. The acquisition time for establishing the search tree is linear and swift.

How to Display

In order to substantiate this method a program, particularly regarding standard graphics hardware acceleration is employed to epitomize real-time motion of the robot, which leads to demonstrating the ability to correct the observer's position. Two preprocessing steps, namely rasterization of the c-space and computation of navigation function, are indispensable in the graphical description. C-space is generated at the commencement of the computation and not until a change has taken place in the position, the geometry of the obstacles or the robot's geometric structure, is there no point in redefining open space. If a change has happened in the target position, the navigation function should be recalculated. As previously described, calculations of the navigation function are linear so that they can be run at a high speed. Although, the quantization corresponding to graphical description can be carried out roughly contemporaneously while monitoring the robot's motion.

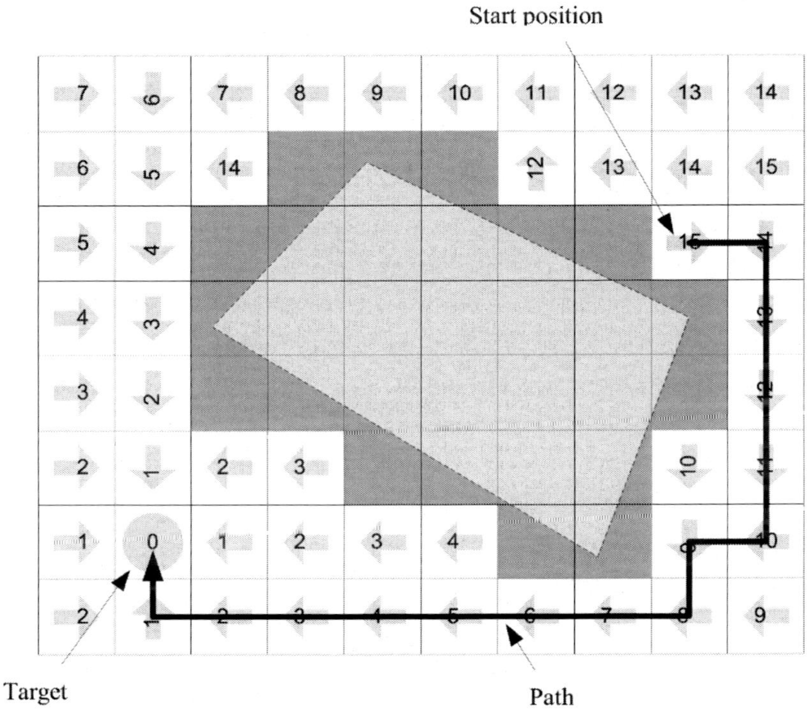

Figure 108. Generating paths in the desired algorithm.

Hardware Utilization

Despite this algorithm speed during running in the non-parallel processor is very low, it proves to be exceedingly high by means of executed hardware or parallelism. The first part (creation of the c-space obstacles), and fourth part (simulation, kinematic) need graphics hardware while running. The second section (flood-like expansion) refers to a local performance and is of optimized computations [65]. The third part (depth initial search tree) is a fast-sequential operation by its nature.

Testing Algorithms

The presented algorithms have been examined in environments with manifold obstacles, moreover, robots and obstacles have different geometric convexities and shapes. The time span for the examination of each samples is drawn up in Table 5. Anytime that stages in preprocessing for configuration space, and navigation function are finalized, the path planning algorithm is implemented in real time. Considering this, the real-time movement will be attainable. Processing, and the achieved results are executed on a *HP 835* system with *turbo-SRX*. Pre-calculations have been performed for Figure 109 and Figure 110 (Marked with * in Table 5) remarkably similar to Figure 112, and duplication is not required.

Example 1: Two-Dimensional Robot Motion in $R^2 \times S^1$ Space

In this instance, the motion of a two-dimensional robot with three degrees of freedom (two translation and one rotation) is going to be inspected. The robot under examination and its physical workspace have been demonstrated in Figure 109, according to which encountering rectangular and triangular obstacles on the way of the robot are affixed to the in-between walls as well as two other small triangular barriers in the top left corner. The position of the light gray target is tagged with the letter (d), and three occasions of (a), (b) and (c) are also earmarked for the beginning position of the robot in order for the algorithm performance to be evaluated as can been depicted in Figure 109.

Table 5. Run time for each example

Problem	Figure	Pixel netting	Run time			
			c space	Fill	Search tree	Display
two-dimensional robot with direct path traversing	110	256 × 256 × 120	22.6	44.1	0.11	35.8
two-dimensional robot with reverse path traversing	111	256 × 256 × 120	*	*	0.11	50.3
trammeled two-dimensional robot	112	256 × 256 × 120	*	*	0.01	0
piano (environment 1)	116	192 × 192 × 180	26.0	50.2	0.05	27.1
piano (environment 2)	117	92 × 92 × 90	8.1	8.1	0.04	7.2

The culmination of the navigated path for the starting point (a) has been portrayed in Figure 110. Carefully scan this figure, it can be seen that after a small number of initial steps of the robot, it manages a 45-degree clockwise rotation in the top right corner of the environment, and subsequently, retains its position nearly up to its reaching to the small triangle obstacles of the upper left corner.

The robot, with the aim of traversing triangular obstacles, initiates performing another counter-clockwise movement and persists its motion as much as 180° rotation so as to reach the target position. As is noticeable, all these traveling routes are carried out with no stepping backward or returning. As a matter of fact, since the corresponding algorithm sets off its path through looking for from the target side and proceeds until the starting status, hence it is capable for detection of the available path (s), and afterward, selection of one which has the best degree of efficiency.

In the starting position (b), the robot gets into a gripping situation. As is evident from Figure 111, in order for the robot to go towards the target point, it is supposed to pass through triangular obstacles, and doing so entails rotating counterclockwise.

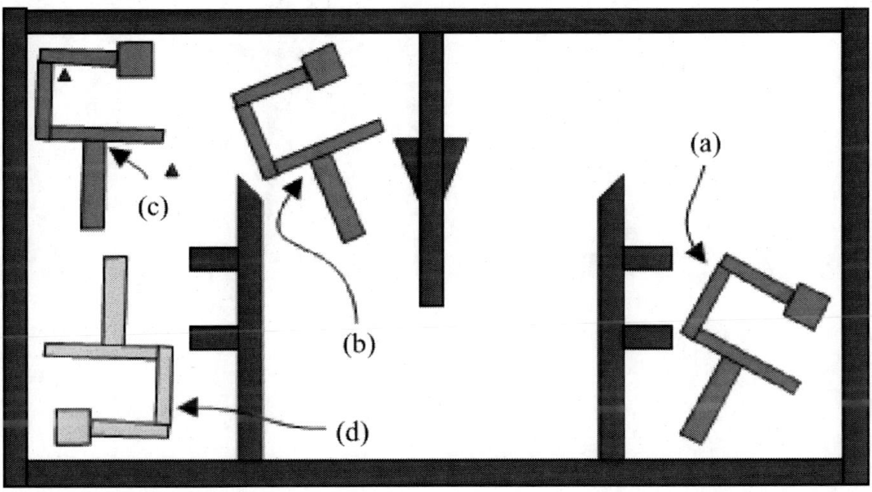

Figure 109. Robot's positions.

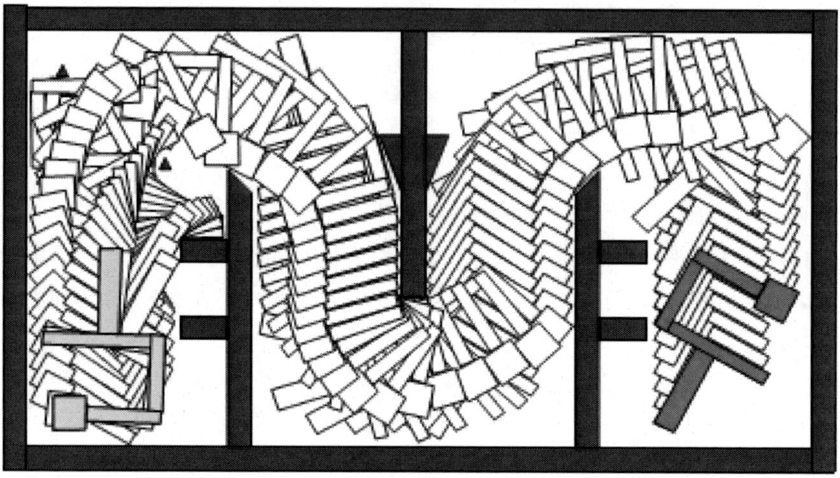

Figure 110. Mobile two-dimensional robot without returning to the target.

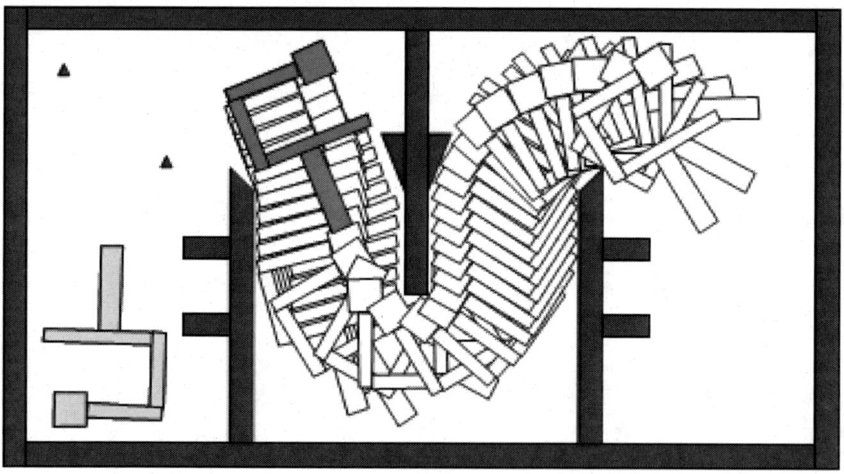

Figure 111. The first step to navigate the robot towards the target in mode (b) (reverse scroll to correct the situation).

As can be reflected, the positioning state of robot, and obstacles is, nevertheless, in a way as to ensure that there is no prospect for rotation at the starting point. On that ground, the algorithm steers the robot towards a reverse path in its initial move unless it approaches the top right corner place of the environment where there exists a higher likelihood of rotating.

Thereupon, in the second step (Figure 112), the necessary rotation puts the robot in a situation akin to state (a). As a result, it is able to keep on its path towards the target position. The algorithm doesn't keep itself within the bounds of the path from the starting position to the target, this characteristic takes into account as a critical feature without which there exists no paths towards the target (Figure 113).

Nonetheless, the conditions are totally dissimilar in location (c). As shown in Figure 114, the robot situation is in such a way that the posing triangular obstacles intercept the robot from performing any maneuver. Eventually, there exists no path towards the target location (c).

Example 2: Three-Dimensional Robot Motion (Piano) in $R^2 \times S^1$ Space

Plenty of three-dimensional robots have a variety of components differing vertically from one another. By way of illustration, the piano in Figure 115 possesses small bases and the large main part. If two kind of short and long obstacles are defined, they can be distinguishable. The short obstacles impede the motion of bases, and long obstacles restrict the body of the piano. Consequently, in such circumstances, this algorithm which is basically devised for two-dimensional issues turns out to be operated in three-dimensional problems, which can be divided into two or more two-dimensional ones.

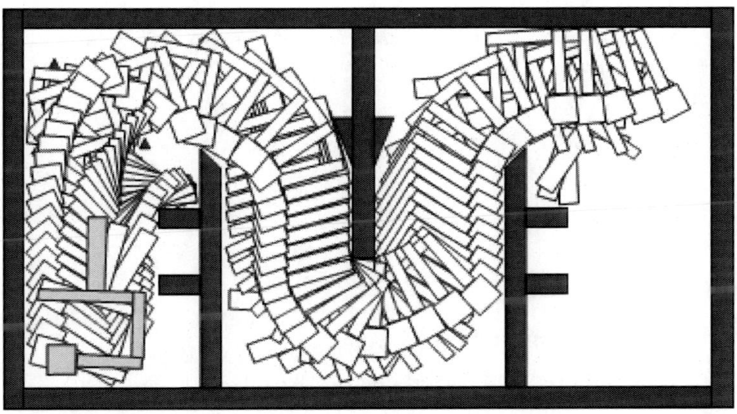

Figure 112. The second step to steer the two-dimensional robot toward the target in state (b) (state correction and direct scrolling).

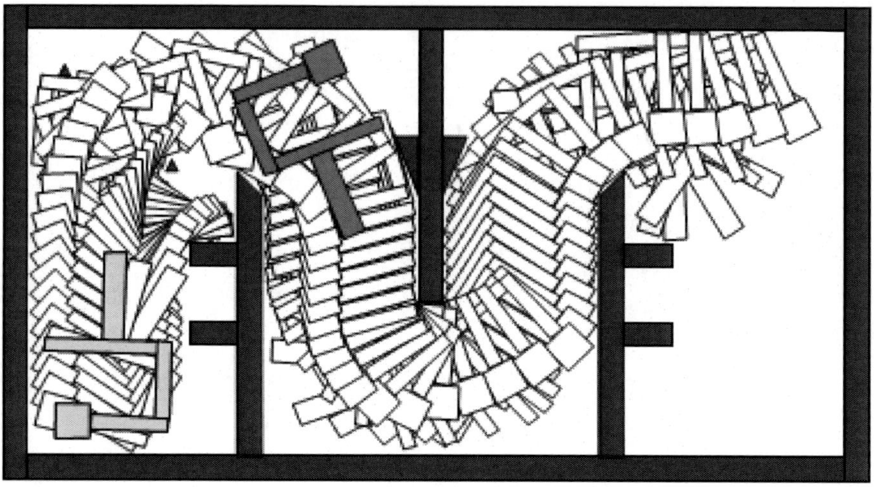

Figure 113. Combining the first and second steps to move towards the target in mode (b) (inverse and direct scrolling).

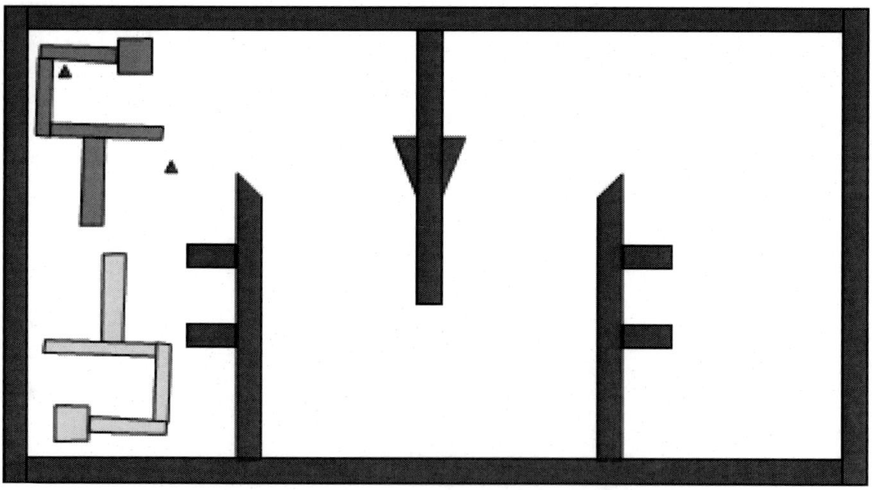

Figure 114. The trapped two-dimensional robot in mode (c).

Rasterizing the environment is made just same as before, with the exception of c-space creation in the case under consideration, the body and bases will be summed together through Minkowski approach with big obstacles and the other low obstacles, respectively. Accordingly, a single bitmap with rasterized c-space obstacles will be specified by means of the union of high obstacles and low ones. Given that the summation of short obstacles and the piano's body did not calculate Minkowski sum, the body would pass through these obstacles; however, the bases can solely perform maneuver around them.

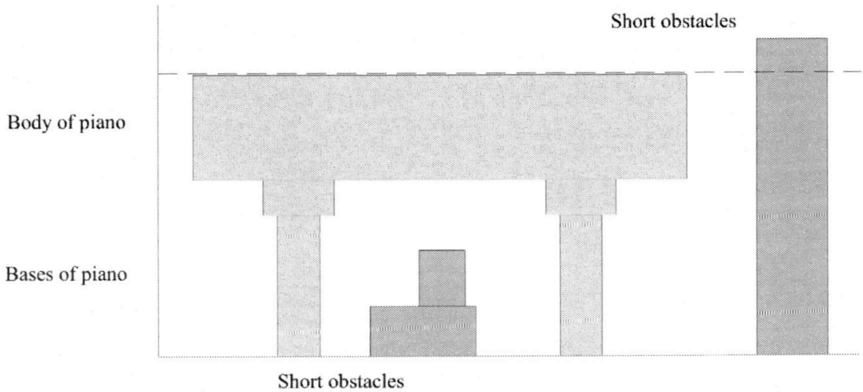

Figure 115. The piano's division into two types of bases and body.

Experiment on piano movement was carried out in two separate environments in each which two different starting points were taken. As shown in (Figure 116), the first environment has both long and short obstacles (body and base), in which long and short obstacles are pinpointed as dark and white squares, respectively. Furthermore, the positions of bases from the top view are distinguished as shapes with three small squares on it. As evidenced, the piano from both starting points moves through the direct paths and rotates if needed while bypassing short and high obstacles in order to achieve the target position.

Only short obstacles (bases) exist in the second environment (Figure 117). There happen two routes towards the target in the initial beginning point (Figure 117a), one of which is the direct path till the end of the aisle and another which is straight direction towards the left side of the exit. Turning to the issue of the path selection using algorithm under consideration which is based upon minimizing rotation, the first path (undeviating movement towards the end of the aisle) is chosen. In the second starting point (Figure 117b) piano also selects the straight path and makes required rotation aligned with the target position in the end route.

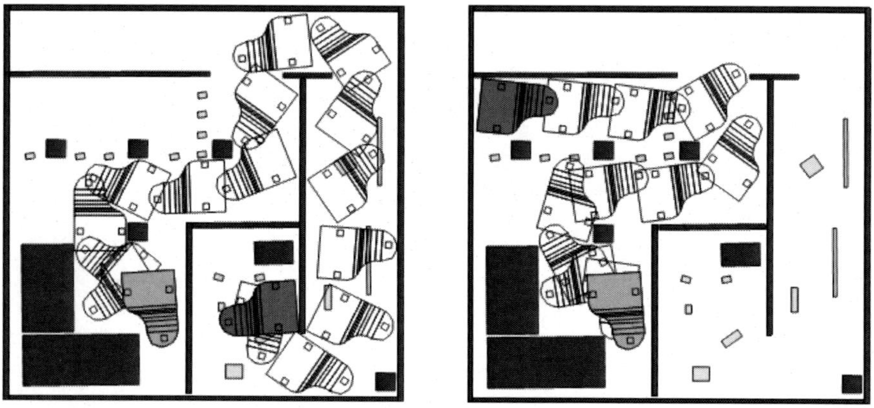

(a) (b)

Figure 116. The piano's movement in the first environment, comprising of both short and long obstacles, from two different starting points.

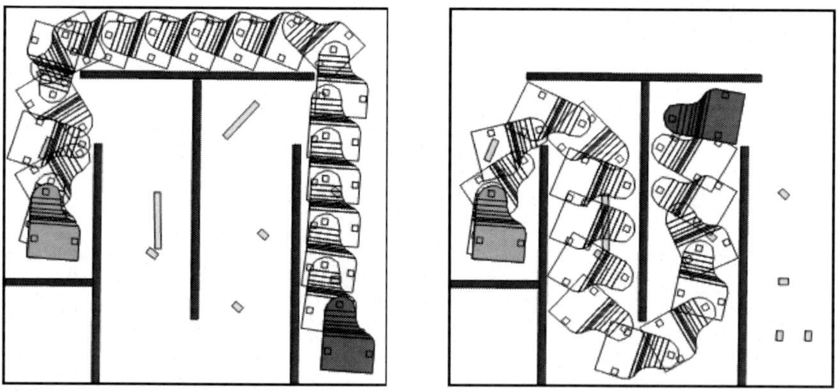

(a) (b)

Figure 117. The piano's movement in the second environment, including short obstacles, from two different starting points.

Chapter 5

IMPLEMENTATION OF PATH PLANNING ALGORITHMS

INTRODUCTION

In chapters 2 to 4, different kinds of path planning algorithms and methodologies in both known and unknown environments and besides the due way to rely upon the audio and visual sensors were introduced, and also their operating principles were delineated in considerable detail.

In this chapter, implementing a number of the foremost strategies will be advanced. The second part of the chapter is dedicated to the implementation of Bug 1, Bug 2, and DistBug algorithms on the Pioneer Robot. On top of that, their traversed path and experimental virtue of these methods are collated [68]. In the third section, an analytical approach presented in order for the performance of the robot to foster in such a way that collision with obstacles is avoided. This method aims to proceed the robot through assorted convex and concave obstacles (obstacles whose shapes are Z, V, U and zigzag), on account of tangential escape [69]. Setting up paths within the environments with and without risk space has proceeded in the fourth and fifth sections of this chapter. As is well known, there could be three sorts of areas in the robot's paths: free, risk and obstacle. The robot is supposed to travel through open spaces but avoids obstacles. Decisions regarding risk areas (incl. muddy paths and water holes) directly depend upon the availability of chosen options. The line space can be the sole path to cross, or possibly there may be several other paths through which to move. Four approaches, namely artificial fuzzy potential field, lingual method, Markov decision process, and fuzzy Markov decision process, are presented in these two parts and tend to be implemented on Nao humanoid robot. By way of conclusion, overall performance results of these procedures are going to display as images generated from crossed paths and the robot decisions once experienced with obstacles.

Comparisons among Bug Family Algorithms

In this section, path planning algorithms including Bug 1, Bug 2, and DistBug will be implemented on mobile robots, and subsequently, their obtained results are going to be analogized. Path planning through referenced above algorithms is imposed on a Pioneer robot in a very simulated environment. In these examinations, acoustic sensors are utilized.

As stated in the second chapter, Bug algorithms as straightforward approaches guarantee the high performance, based on which if the robot encounters obstacles, they will be ready to take a path towards the target (if available). In this class of algorithms, three general assumptions about robots are made: 1) Robot is considered as a point body. 2) The robot is equipped with complete localized information from its environment. 3) The robot's sensors are accurate.

Implementation

The codes for Bug algorithms are frequently established in multiple references. In this survey, the associated codes are written in C++ and ran using the Aria libraries in the Linux environment. The simulation environment has 10 meters length, 8 meters width and two convex obstacles (Figure 118). To draw a map, Mapper ver.3 explained in reference [49] is utilized.

The implementation of algorithms, during which the robot updates its positioning and navigation information making use of acoustic sensors, enables the robot to detect boundaries (or walls) and cross towards the target by virtue of locating the corners of unknown obstacles.

In the simulation setting, the robot utilizes a distance meter so that it is possible to spot and update its information (incl. position and direction of the traveled path). After instantaneously updating the status of the robot, it is able to manage the collisions with obstacles and recognize their leaving points. Into the bargain, it is also vital for the robot to own the capacity of maintaining itself traveling along a line with a gentle slope at any given time during the implementation of the Bug 2 algorithm. It is worth noting that despite the position of the robot seems to vividly visible within the simulated environment, the full position in a real environment is hardly accessible.

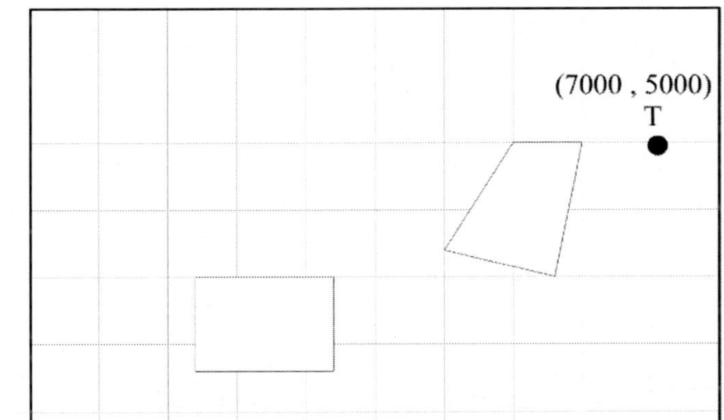

Figure 118. Simulated environment in the MobileSim Software to assess Bug 1, Bug 2, and DistBug algorithms.

Identifying obstacles and tracing their border by means of acoustic sensors is appeared in Figure 119. The robot employs frontal sensors (marked with numbers from 2 to 7) to distinguish obstacles facing it. It likewise utilized sensors number 8 and 9 in order to follow the obstacle's boundary in a clockwise direction and again, sensors numbers 1 and 16 in a counter-clockwise direction. Sensors numbers 10 through 15 are linked to reverse movement of the robot.

Tests and Results

The robot is situated on a two-dimensional environment whose coordinates of the starting point and target point are $(0,0)$ and $(5000,7000)$, respectively. The length of the shortest path between the intersection of the path from the starting points (S) to the target point (T) is 8,602 mm, and the robot's speed is 100 millimeters per second (mm/s). The results of the execution of Bug 1, Bug 2, and DistBug algorithms are displayed in Figure 120.

Figure 121 also portrays the graphs of traveled length using three algorithms of Bug 1, Bug 2, and DistBug in line with the path of environment-dominant bird (the best possible path). Refined results applying Bug 1, Bug 2, and DistBug algorithms are clear-cut in this figure.

As manifested in Figure 120a, in Bug 1 algorithm, the robot scans itself crossing from all the borders of the faced obstacles, during which the robot has the slowest performance in stark contrast to the other two strategies. Even with this inconvenience, this algorithm guarantees reaching the target. As a result, the length of the navigated path by the robot executing this method is 29,022 mm.

Utilizing Bug 2 algorithm (Figure 120b) once the robot confronted with the first obstacle and acknowledged an initial slope, it transcends the obstacle from the left side. Only then does it again cut the line connecting the starting point to the target point; it proceeds the line on the same route. For the next obstacle, this procedure is also duplicated. As shown in the figure, unlike Bug 1, the Bug 2 algorithm does not fully transcend the barrier, therefore as a consequence, it speeds up reaching the target. With this respect, the length of the traversed path by the robot using Bug algorithm 2 is 16,194 mm. All the same, in a tortuous route these two algorithms do not have any overwhelming superiorities over one another.

Whatever the case, the traversed path in DistBug algorithm (Figure 120c) surpasses Bug 1 and Bug 2 altogether. This algorithm accompanying the capability of the sensors makes detecting the shortest route possible to the most significant degree. Literally, while the robot trying to find associate obstacle's boundary, it consistently measures the direction of its current position to the target and oversteps the obstacle just as soon as noticing its free path towards the target. That is why the shortest travelled path among the three studied algorithms belongs to DistBug. Not only that but DistBug also is able to fulfill constraints referred to simultaneous movements towards the target while tracking the obstacle's boundary. Ultimately, the traversed length of the path by the robot implementing this algorithm is 12,795 mm.

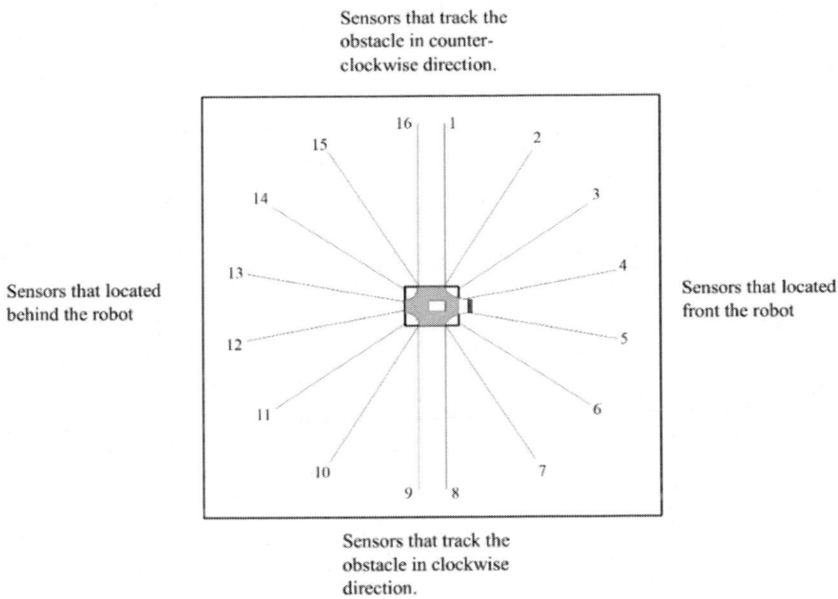

Figure 119. The positioning of the robot sensors implemented in the evaluation of Bug 1, Bug 2 and DistBug algorithms.

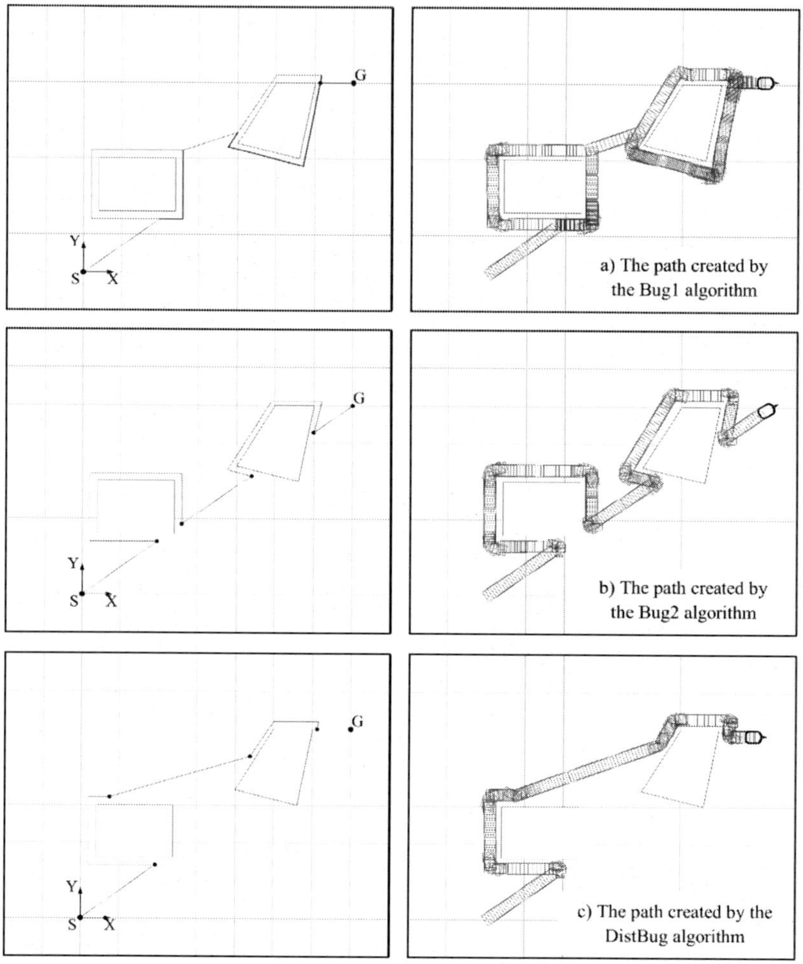

Figure 120. The traveled routes generated by Bug 1 (up), Bug 2 (middle), and DistBug (down) algorithms.

An Analytical Method to Avoid Collision of Mobile Robot with Obstacles

In this section, to start with, the modelling relations of searching for the target at open space, involving kinematic model, position control, rotation control and the efficient way for stabilizing during changing control systems are proposed. At that point, based on this modelling, a stable non-linear controller is provided amid movement for dodging obstacles ahead that may collide with the robot while proceeding towards the target. At whatever point the robot detects an object quite close itself, the rotation angle alters with respect to the tangent of the angle between the robot and the corresponding object. Thereupon, the robot takes steps to look for a path to reach the target by following the obstacle. At last instance, simulation and application results of utilizing this technique on the robot are put forward.

Target Searching at Open Spaces

Through this strategy, the robot carries out its mission via the subsequent two tasks

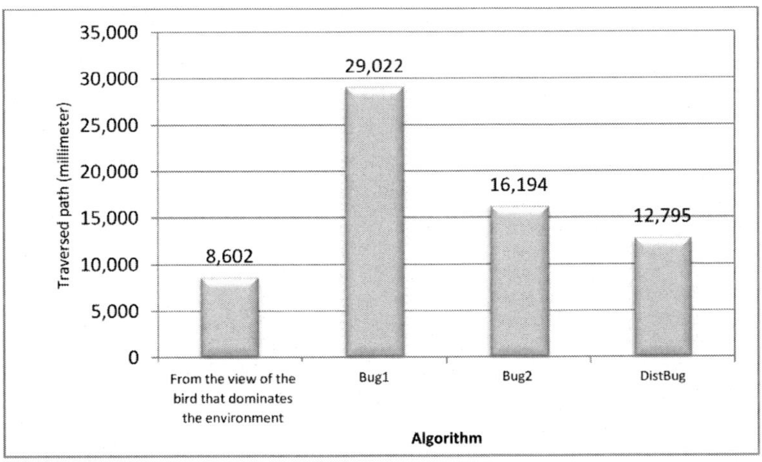

Figure 121. The length comparison of the robot's traveled routes applying the diverse routing algorithms.

First task: shortening the distance to the target in a case that there is no obstacle ahead of the robot within a half-circle with a diameter of d_{obs}. In such a case, the rotation of the robot is permitted to keep away the collision of the robot with the obstacles

Second task: The first task is performed again after going away from the obstacle so as to decrease the distance to the target steadily. Subsequently, the robot reaches to the target through overtaking each of the obstacles. Other than, on condition that the robot requires to rotate after locating the target, it is conceivable to do so coupled with maintaining its position. In Figure 122, The block diagram approach to this process is illustrated.

Kinematic Model of the Robot

In the current experiment, the differential equation that is dominated on the robot is defined as follows:

$$\begin{bmatrix} \dot{x} \\ \dot{y} \\ \dot{\psi} \end{bmatrix} = \begin{bmatrix} cos\psi & -asin\psi \\ sin\psi & acos\psi \\ 0 & 1 \end{bmatrix} \begin{bmatrix} u \\ \omega \end{bmatrix} \tag{24}$$

in this equation, u and ω represent linear and angular velocities, respectively. Moreover, x and y are regarded as current coordinates of the robot, and ψ refers to the deviation extent from axis x in world coordinate system $\langle O \rangle$, and finally, a is a distance from where the controllability point to the center axis is connected to the flywheel (Figure 123).

In an environment with open obstacles, the mathematical representation of the robot navigation to the target is scrutinized in the polar coordinates system [70]:

$$\dot{H} = \begin{bmatrix} \dot{\rho} \\ \dot{\alpha} \\ \dot{\theta} \end{bmatrix} = \begin{bmatrix} -cos\alpha & -asin\alpha \\ \frac{sin\alpha}{\rho} & -a\frac{cos\alpha}{\rho} - 1 \\ \frac{sin\alpha}{\rho} & -a\frac{cos\alpha}{\rho} \end{bmatrix} \begin{bmatrix} u \\ \omega \end{bmatrix} + \begin{bmatrix} \dot{x}_d cos\theta + \dot{y}_d sin\theta \\ -\dot{x}_d sin\theta + \dot{y}_d cos\theta \\ \dot{x}_d sin\theta + \dot{y}_d cos\theta \end{bmatrix} = \boldsymbol{Kv} \tag{25}$$

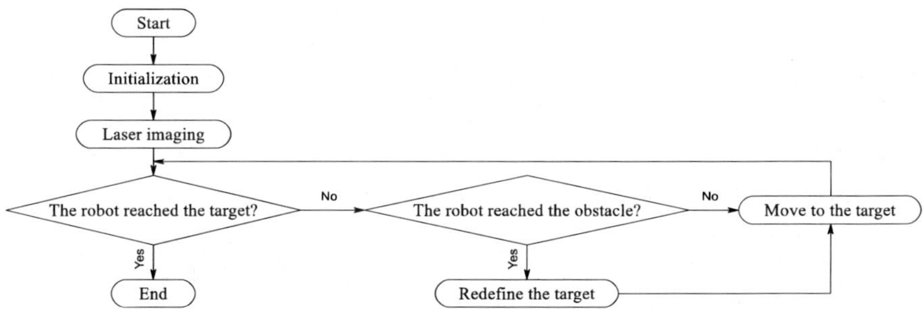

Figure 122. The proposed process block diagram.

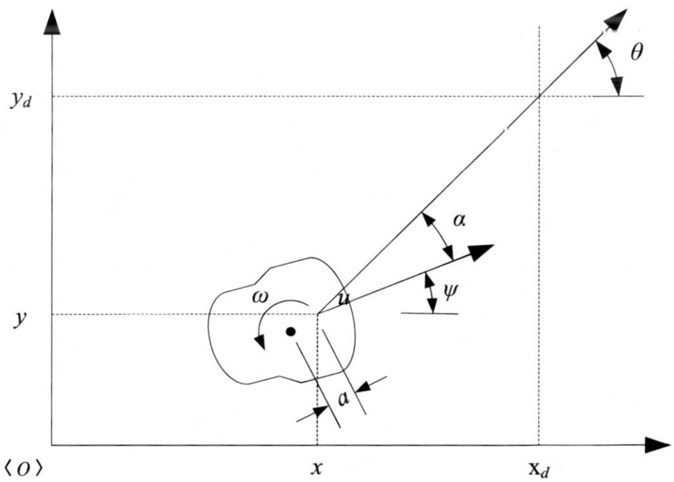

Figure 123. The variables of the movement of a robot towards the target.

where, ρ denotes the robot distance to the target. Should the target be static (notably: $\dot{x}_d = \dot{y}_d = 0$), this relation is summarized as:

$$\dot{H} = \boldsymbol{K}\boldsymbol{v} \tag{26}$$

in which, $\boldsymbol{K}$ defines the first matrix in the equation (25), v specifies linear velocity vector, and angular velocity vector as $[u \quad \omega]^T$. It should be underlined that ρ cannot have zero value, since in the case under consideration, both α and θ will be undefined. Accordingly, with the estimation of $\rho \leq \delta$, it may be presumed that the robot places in the target position where it is considered δ = *30mm.*

In Figure 123, the target is presented with $\langle t \rangle$, and its coordinates in $\langle O \rangle$ are characterized by the user. The direction of motion is invariably perpendicular to the virtual axis of the robot and draws angle α with the target's position. The angle α is determined as the rotation angle of the robot at the very moment of its arrival in the target point and can be yield by the succeeding equation:

$$\theta = \alpha + \psi \tag{27}$$

Position Control in Open Space

From Figure 123, it can be figured out that the robot can be effectively controlled implementing efficient and proper control of u and ω values. Control objectives in this part are taken as $\rho \rightarrow 0$ and $\alpha \rightarrow 0$. Furthermore, the condition $\theta \rightarrow \theta_d$ is supplemented to both before-mentioned conditions, further in this section. Referred herein to the same degree, the objective is to have the position vector $H = [\rho \quad \alpha]^T$ approached to zero under optional initial conditions. Considering this, as a means to design a controller complying with the mentioned conditions, Lyapunov function is defined by the next equation as:

$$V = \frac{1}{2}\boldsymbol{H}^T\boldsymbol{H} > \boldsymbol{0} \tag{28}$$

For closed-loop system stability $\boldsymbol{H} \rightarrow [0 \quad 0]^T$. It is crucially essential for the derivation of this function to get negative by-product for all non-zero values $\boldsymbol{H}$. The substitution of output from equation (28) for (25), it can be derived:

$$\dot{V} = \frac{1}{2}\dot{\boldsymbol{H}}^T\boldsymbol{K}\boldsymbol{v} > \boldsymbol{0} \tag{29}$$

where $\boldsymbol{H}$ has a negative value for each non-zero amount. It is, therefore, desirable that the control rule is defined as follows:

$$v = \boldsymbol{K}^{-1}(\dot{\boldsymbol{H}}_d - \kappa\ tanh\boldsymbol{H}) \tag{30}$$

in which, $\boldsymbol{\kappa}$ indicates a known particular diagonal matrix. Therefore, taking into consideration of $\dot{\boldsymbol{H}}_d = 0$ (for static target) it leads to:

$$\dot{V} = -\boldsymbol{H}^{\boldsymbol{T}}\kappa\ tanh\boldsymbol{H} < \boldsymbol{0} \tag{31}$$

when → 0, this implies that the placement of the robot at the target point during the absence of obstacles, is guaranteed. In case of robot arrival to the target, the values u and ω become zero, which allude to the robot staying static as soon as the robot reaches the target position. After fulfilment of the condition $\dot{\boldsymbol{H}}_d = 0$ in the equation (30), the following equation can be achieved:

$$\begin{bmatrix} u \\ \omega \end{bmatrix} = \begin{bmatrix} u_{max} & 0 \\ 0 & \omega_{max} \end{bmatrix} \begin{bmatrix} \frac{\rho + a cos\alpha}{a + \rho cos\alpha} & \frac{\alpha\rho cos\alpha}{\alpha + \rho cos\alpha} \\ \frac{sin\alpha}{a + \rho cos\alpha} & 1 - \frac{a}{a + \rho cos\alpha} \end{bmatrix} \begin{bmatrix} tanh\rho \\ tanh\alpha \end{bmatrix} \tag{32}$$

In order for the stability analysis to be thoroughly evaluated, the angular behaviour of θ in terms of time once the robot arrives at the target point, needs to be scrutinized. With the enlightenment that $\dot{\psi} = \omega$ and the system also has asymptotic convergence, with the robot accosting to target, ω also tends to zero ($\omega \to 0$). On the contrary, by having $\rho \to 0$ results in $h\,\rho/\rho \to 0$, which subjects to $t \to \infty$, it tends to be 0. Hence, $\theta = \psi$ converges to a specific value, which betokens a sound performance of the robot to attain the target. Function $tanh$ is employed in order to maximize values u and ω. Amounts u_{max} and ω_{max} can be drawn out from the robot's catalogue.

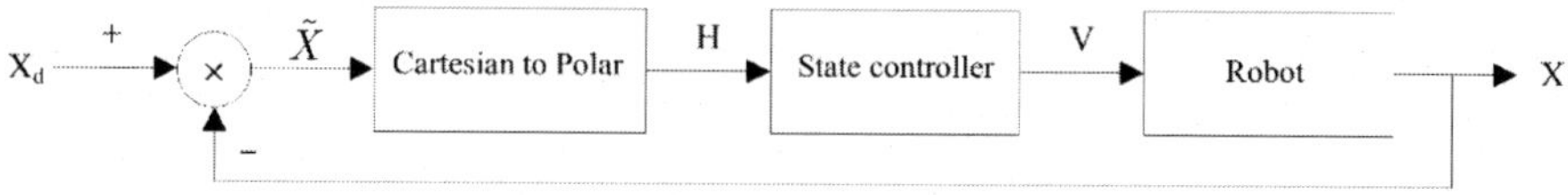

Figure 124. Robot control system moving towards the target in an unobstructed environment.

The suggested control system for robot guidance towards the obstacle-free environment is illustrated in Figure 124. $X_d = [x_d \quad y_d]^T$ denotes the target position and $X = [x \quad y]^T$ is the robot's current location that can be extracted from the traversed path. In addition, it can be obtained:

$$\tilde{X} = [\tilde{x} \quad \tilde{y}]^T = [(x_d - x) \quad (y_d - y)]^T$$

where

$$\sqrt{\tilde{x}^2 + \tilde{y}^2 = \rho} \text{ and } tan^{-1}(\tilde{y}/\tilde{x}) = \theta.$$

Rotation Control at Target

Rotation for the robot is sometimes needed to be done after coming to the target position. For this objective, angle ψ ought to be solely controlled without taking any action. To accomplish this aim, an extra controller is planned in an effort to implement the required angular velocity (ω) to the robot. Meanwhile, its linear velocity (u) is still zero. As an instance, in Figure 123, $\theta = \psi_d$. So, the final rotation error is defined as $= \psi_d - \psi = \tilde{\psi}$. In this way, for a constant last rotation, $\dot{\alpha} = -\dot{\psi} = \dot{\tilde{\psi}}$ remains constant. In a case like this, Lyapunov function can be taken as follows:

$$V(\alpha) = \frac{1}{2}\alpha^2 \tag{33}$$

where the first derivation of this equation with respect to time is as follows:

$$\dot{V}(\alpha) = \alpha\dot{\alpha} = \tilde{\psi}(-\omega + u\frac{\sin\tilde{\psi}}{\rho} - \alpha\omega\frac{\cos\tilde{\psi}}{\rho} \tag{34}$$

Given that:

$$\begin{cases} u = 0 \\ \omega = \omega_{max}\tanh\tilde{\psi}\ (\omega_{max} > 0) \end{cases} \tag{35}$$

It should be emphasized that constraints happen at the target position ($\rho \leq \delta$). Therefore, the following formula is acquired:

$$\dot{V}(\alpha) = -\omega_{max}\tilde{\psi}\tanh\psi\,(1 + \frac{u}{\delta}\cos\tilde{\psi}) \tag{36}$$

This statement is definite negative conceding that $|\tilde{\psi}| < {}^{\pi}/_{2}$. Necessarily, the rotation error ($\tilde{\psi}$) proceeds asymptotically to zero, which signifies ψ tends to ψ_d.

Stability in Switching Control System

Subsequently, after the robot's arrival to the target, it would have to position itself in an excellent final state by means of rotating. Indeed, it should be paid heed to switch from positional to the situational controller to get into a fitting situation in connection with the robot placement direction. In such a situation, the stability of the control system needs to be guaranteed in the course of switching the controller that can be accomplished with Lyapunov's theorem expansion [71]. Considering pioneering studies have proposed that Lyapunov's functions are identical, needless to say, this condition is well stablished on its own accord and stability is assured whilst switching from the positional controller to situational one.

Suggested Method

In this section, a nonlinear controller with a focus on the guidance of the robot without colliding the obstacles is presented. A strategy has contributed to the choice of a tangential direction on obstacles' boundary. This strategy is strictly comparable to guiding the robot through unknown environments.

The controller utilizes recommended techniques in the fourth section so as to shorten the length of the robot path towards the target. The robot perpetually redefines the target position and approaches it in advance of reaching the distance $\boldsymbol{d_{obs}}$ from the obstacle. Nevertheless, when the robot comes to distance d_{obs}, as evidenced in Figure 125, takes aggressive action.

As demonstrated in Figure 125, if the robot comes across an obstacle (if so $\boldsymbol{d_{min} \leq d_{obs}}$) hypothetical target ($\boldsymbol{X_v}$) is procured owing to the implementation of a rotation matrix to the real position of the target ($\boldsymbol{X_d}$). It should be stressed that the line linking controlled points and the virtual target is considered to be tangent on the perimeter.

Rotation angle $\boldsymbol{\gamma}$ specifying the rotation matrix could be measured by dint of a set of quantification of range-meter sensors. It is imperative to define $\boldsymbol{\gamma}$ for taking cognizance of β that ascertains the position of the closest obstacle. After every turn of computing range, the presented system determines the distance d with the proviso that $\boldsymbol{d \leq d_{obs}}$ and the angle β is contingent upon it. The applied range-meter sensors enclose the range $[0^\circ \quad 180^\circ]$ with 1°per step. Further to this, interval $[-90^\circ \quad +90^\circ]$ can be spanned with a range of 180°. On this premise, as long as the obstacle is placed at the right side of the robot, $\beta < 0$ and otherwise, $\beta > 0$. By predetermined β, the rotation angle can be defined as:

$$\gamma = \begin{cases} -90^\circ + \beta - \alpha & \text{if } \beta \geq 0 \\ +90^\circ + \beta - \alpha & if\ \beta < 0 \end{cases} \tag{37}$$

if $\alpha > 0$ it indicates that the target is located at the right side of robot axis movement. The rotation angle γ in Figure 125 has a positive value and triggers the rotation of real target towards the left side.

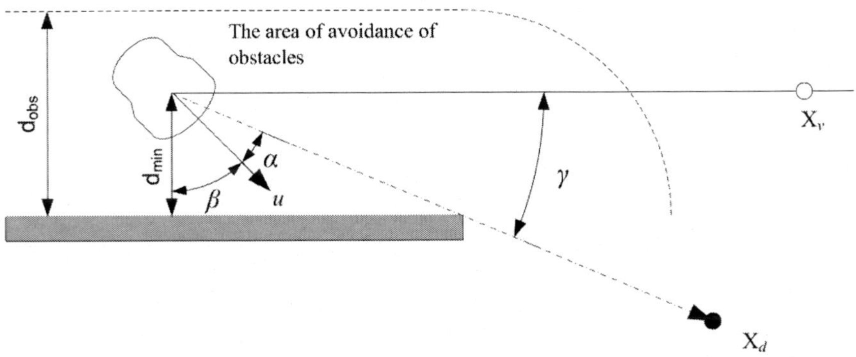

Figure 125. Tangential escape strategy function adjacent to an obstacle.

Recognizing the rotation angle γ, hypothetical and real target positions can establish a connection with each other as follows:

$$\boldsymbol{X}_v = \begin{bmatrix} cos\gamma & -sin\gamma \\ sin\gamma & cos\gamma \end{bmatrix} \boldsymbol{X}_d \tag{38}$$

in this equation, $\boldsymbol{X}_d$ and $\boldsymbol{X}_v$ stand for the position of the virtual and actual targets. Positional controller whose coordinates is $\boldsymbol{X}_v$ goes along the obstacle's border tangentially and leads the robot to the target. It is noteworthy that regarding the occasion of non-existent obstacles, the actual target position will be remained unchanged ($\gamma = 0^\circ$) and the robot steadily nears the target.

The block diagram representation of the planned controller is portrayed in Figure 126. As is evident in this diagram, it can be noticed that the situational controller block stays consistently active. Thus, the main implication is that the robot does not stop to curb hitting the obstacles. Referring to this graph, and in furtherance of stabilizing situational controller asymptotically, only then is the robot left behind obstacles completely, it reaches the intended target on condition of target accessibility. In the forthcoming section, discourse on the tangential escape strategy will forge ahead to address a plurality of types of obstacles.

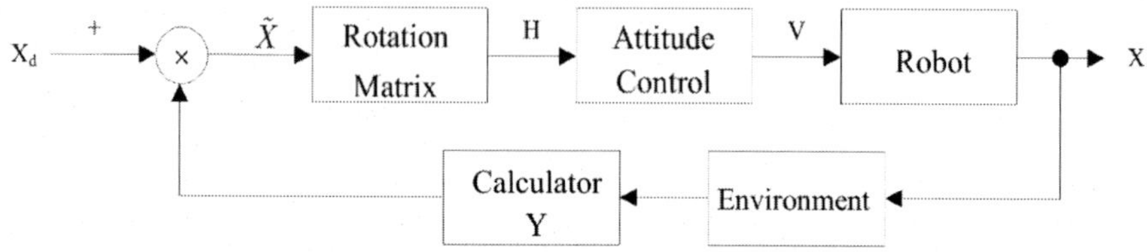

Figure 126. Block diagram of a control system based on tangential scape strategy.

Facing Corner-Shaped Obstacles

The robot occasionally grapples with situations in which attempts more evasive maneuvers to steer clear of obstacles. One of these situations is a confrontation with *V*-, *L*- and *U*-Shaped obstacles. Figure 127 illustrates a status through which the target is spotted under the horizontal segment of an *L*-shaped obstacle.

In this particular case, the rotation angle γ, which is delimited by equation (37), provides no guarantee required for the robot to avoid the obstacle. In practical terms, the robot encounters with the vertical segment of the obstacle and keeps turning around it. In contemplation to expedite this process, the amount γ is compensated once that the distance d_{90} and d_{min} are fewer than d_{obs}, as below:

$$\gamma = \begin{cases} -180 + \beta - \alpha & if\ \beta \geq 0 \\ +180 + \beta - \alpha & if\ \beta < 0 \end{cases} \tag{39}$$

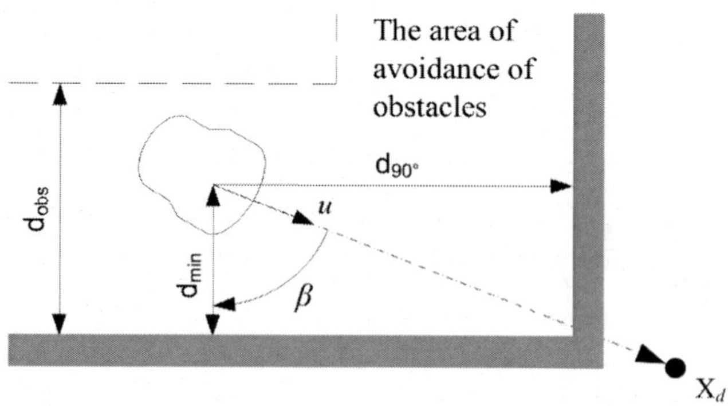

Figure 127. Alternating the tangential escape strategy due to confrontation with the corner-shaped obstacle (L).

Beyond that, the robot, in tandem with its extensive maneuver, desires to transform the definition of the virtual target in the sense that the likelihood of impact with obstacles during the escape maneuver is minimized. This aspiration can be satisfied with reducing the amount of u. A possible solution to diminish u is the reduction of the amount $\rho = \|\boldsymbol{X}_v - \boldsymbol{X}\|^{\frac{1}{2}}$ in the escape operation. According to the invention, the following equation is therefore proposed:

$$\boldsymbol{X}_v = \begin{bmatrix} cos\gamma & -sin\gamma \\ sin\gamma & cos\gamma \end{bmatrix} \begin{bmatrix} d_{min} & cos\theta \\ d_{min} & cos\theta \end{bmatrix} \times \quad (40)$$

In this fashion, the robot starts the process off with a lower linear velocity, ending to reduce the chance of collisions with obstacles during escape tactics.

Facing Concave Obstacles

Another problematic situation requiring a remarkable and independent review is when the rotation of the robot is incongruous with the target. A dramatic illustration of this situation comes into being when the robot has traveled along the edge of the U-shaped obstacle (as can be seen in Figure 128 with the darker field). The robot directly proceeds towards the target from the front of the field and sets foot in the U-shaped obstacle. Only by attaining the depth of the obstacle will the robot follow the border of the obstacle. With regard to Eqs. (38) and (40), the robot reverses to the left by coming close to the upper U-shaped obstacle and moves to the darker position. In this status, the rotation error of α indicates the rotation fault of the robot with respect to the target.

When the robot releases its own from the obstacle, it turns to the left for making progress towards the real target, but it is caught in the U-shaped obstacle, which is regarded as a local minimum trap. In order to be entirely free from these kinds of local minima problems, it is inevitable to consider a new sentence (P_{obs}). After rescanning the environment, if at least one obstacle is detected such that its distance to the robot is less than d_{obs}, the amount of the P_{obs} sentence will be one (1). Otherwise, it will possess the zero (0) as its value. The case in which the robot performs tracing the barrier's border until the end of its mission, one distance, at least, will be identified, whose value is less than the d_{obs}. Accordingly, the amount of P_{obs} remains unchanged as long as the operation is accomplished. Once the robot could cross the obstacle, it applies an approach, which is analogous to the path planning algorithm in anonymous environments comprising of the walls, in order to escape from being trapped. After each time when the value of P_{obs} is changed, the value $|\alpha|$ is reconsidered as well. If a situation in which this parameter is greater than 90, is occurred similar to Figure 128, the robot conducts the maneuver to traverse the route demonstrated by a dash- line in this figure. In such a case, not only is a virtual target produced, but also a tangential scape strategy is implemented to deliver the robot to the target. The coordinated of the new virtual purpose as $[x_{tg} \quad y_{tg}]^T$ is derived from the following relation:

$$\begin{bmatrix} x_{tg} \\ y_{tg} \end{bmatrix} = \begin{cases} \begin{bmatrix} x \\ y \end{bmatrix} + d_{min} \begin{bmatrix} -sin\psi & cos\psi \\ cos\psi & sin\psi \end{bmatrix} \begin{bmatrix} 1 \\ 1 \end{bmatrix} & if\ \beta \geq 0 \\ \begin{bmatrix} x \\ y \end{bmatrix} + d_{min} \begin{bmatrix} sin\psi & cos\psi \\ -cos\psi & sin\psi \end{bmatrix} \begin{bmatrix} 1 \\ 1 \end{bmatrix} & if\beta < 0 \end{cases} \tag{41}$$

in these relations, $\boldsymbol{X} = [x \quad y]^T$ appertains to the robot's current position in the null coordinate system, ψ is the current rotational orientation, d_{min} stands for the shortest distance between the robot and the precursor prior to resetting P_{obs}, and β mentions the corresponding angle with d_{min}. Thereafter, the robot works in the position control mode in order to find the virtual target. Should the obstacle appear in the quest of the virtual target, the robot will neglect and abandon the virtual target given that the real purpose is taken into account. Otherwise, the angular position of the robot is going to be rectified through achieving the virtual target, that being so the robot will locate in the bright gray position as shown in Fig. 128. This is brought about in consequence of using the attitude controller (as outlined earlier). The optimal rotation angle is calculated by applying the following relation:

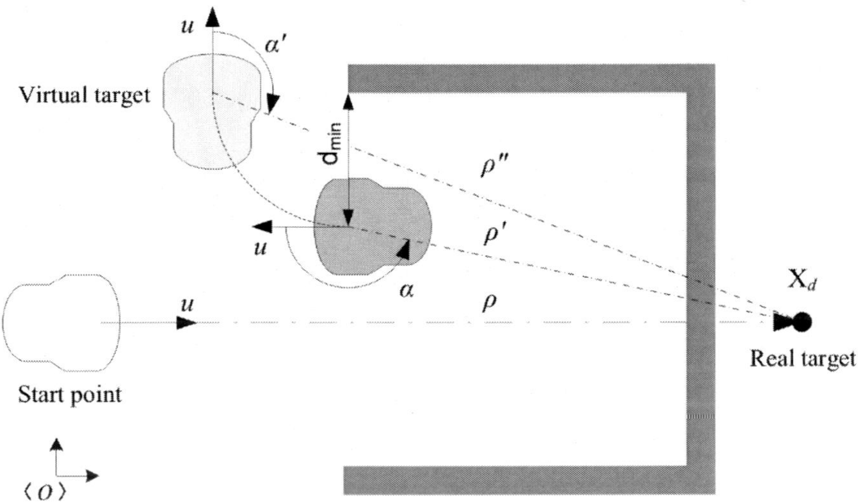

Figure 128. 90-degree rotation of the robot for bypassing the U-shaped obstacles.

$$\psi_{tg} = \begin{cases} \psi - 90^{\circ} & if\ \beta \geq 0 \\ \psi + 90^{\circ} & if\ \beta < 0 \end{cases} \tag{42}$$

in which, ψ is the rotation of the robot on the verge of resetting P_{obs}. After this rotation, the robot is going to be located in the virtual target position.

Using the Stored Information of the Observed Obstacles

Sporadically, the robot might also detect a circumstance within the working environment that it has already been attempting to forestall whilst seeking the target. In the instant case, the robot is going to monitor the obstacle for the second time; this can be interpreted that it enters an endless loop and failures to search out the target.

The planned controller makes use of an intermediate memory to save obstacle enclosures (that is, the point at which the value of P_{obs}increases from zero to 1). In this respect, an impulsive time interval (for instance, 100 milliseconds) is preferred as the reference time interval and associated places accumulated in memory, at the end of every specific interval, are contrasted to the present position of the robot. The robot will keep on moving forward unremittingly if only no commensurate match is sensed (or the memory is empty). In a match agreement situation, the robot will turn one and a half round to discover a route in the opposite direction (utilizing the rotational controller), and stores this point in its mid-term memory and then keeps moving again. The robot is unable to reach the target under any circumstances provided that there are two matches.

Simulation and Laboratory Results

Various actual experimentations and simulations have been conducted applying the control system reviewed with the intention to corroborate the proposed approach, the results of which are exhibited in this section. The examinations were executed on a Pioneer 3D-DX robot (Figure 129).

The robot is appropriately equipped with a laser search engine named Sick LMS100 (installed by the manufacturer). As shown in Figure 130, at each turn, this search engine is able to measure 181 distances, which encompasses a semicircle collectively. The laser scrolling angle is equal to 1 degree, and the sampling rate is $10\,Hz$ (similar to the control systems using in the lower level robots).

In Figure 131, a zigzag-shaped obstacle is examined and afterward the robot path towards the target is demonstrated. During this simulation, the robot experiences six obstacles that have been scanned only once.

Figure 129. The pioneer 3D-DX robot.

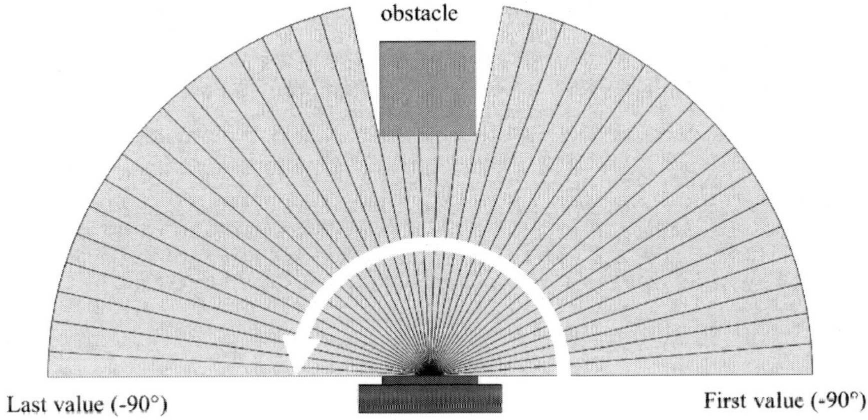

Figure 130. The illustration of storing a range in the Sick LMS100 laser search engine in a semicircle.

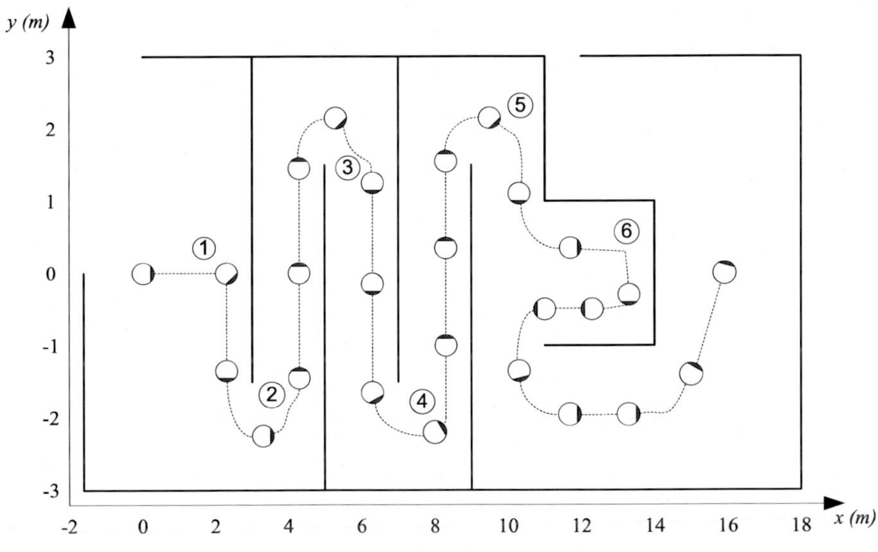

Figure 131. The outcome of the execution of a tangential escape algorithm in the zigzag-shaped environment.

In Figures 132 and 133, black circles indicate positions over which the robot has traversed a couple of times. By retrieving these points, the robot can avail itself of employing the angular position controller as a tool to modify the direction of motion in the opposite direction. In Figure 132, this maneuver is carried out at two points whose coordinates are $(2.5,2)$ and $(5.5,2)$ and in Figure 133 at a point with coordinates $(1.5,0)$. It is incontrovertibly axiomatic that the planned strategy can successfully take the robot to its expected target.

In the following, four other experiments will be presented. In the first examination, with the utilization of the suggested algorithm, the robot can reach the target located behind the U-shaped obstacle, as shown in Figure 134. Position error, angular position error, and rotation of the robot are detailed from Figure 135 to Figure 137. The control signals u and ω are depicted in Figures 138 and 139.

The paths navigated by the robot within the scope of three other examinations appear in Figure 140 to Figure 142. The first case includes a Z-shaped obstacle, inside which there are a variety of different obstacles. The second and third experiments have to do with obstacles in the form of L and V shapes. The results demonstrate that the tangential escape route is competent enough in steering the robot towards the target position and staying away from the obstacles.

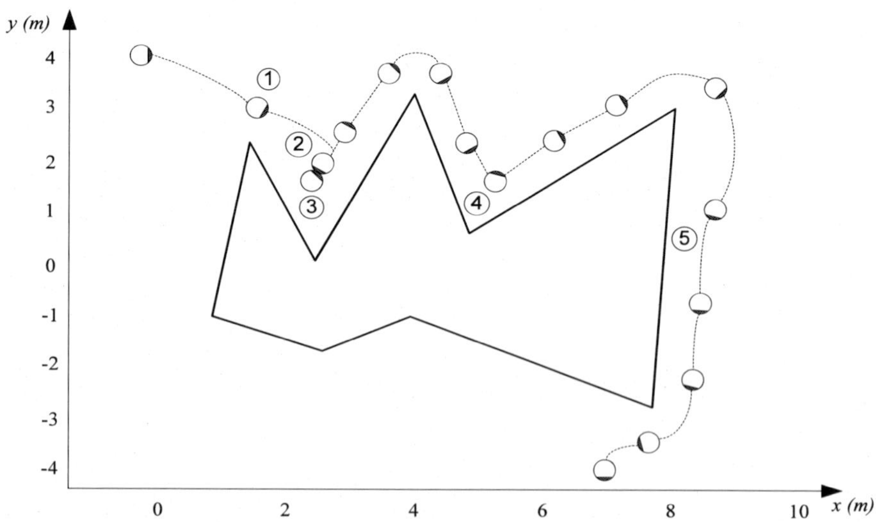

Figure 132. The result of the execution of a tangential escape algorithm in an environment with uneven convex and concave obstacles.

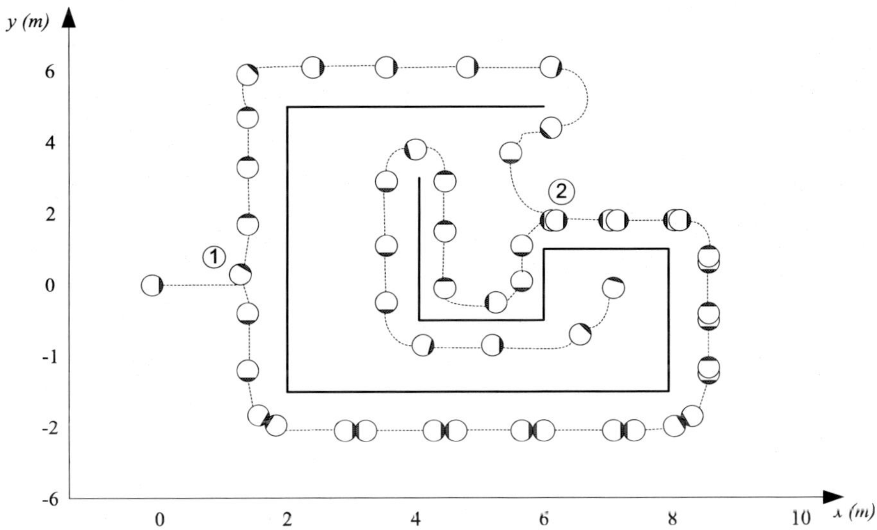

Figure 133. The result of the execution of the tangential escape algorithm in the sample environment.

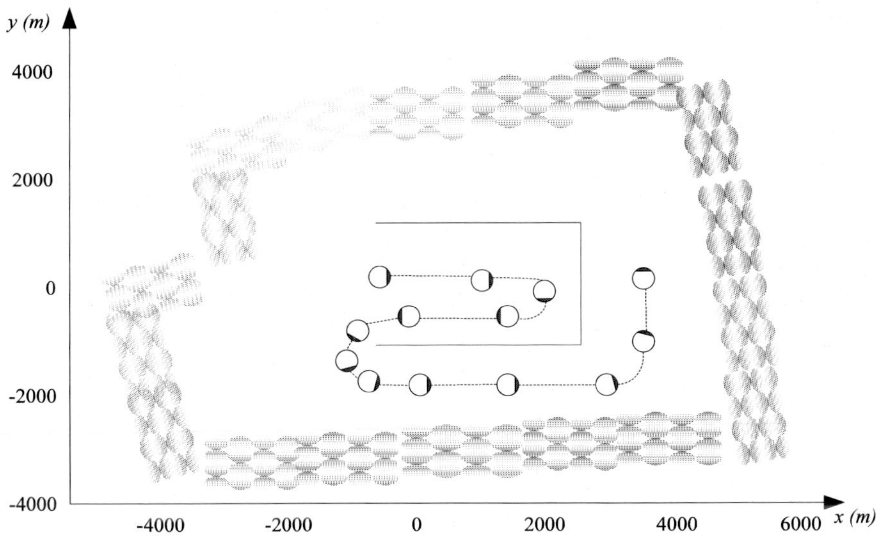

Figure 134. The result of the execution of a tangential escape algorithm in a U-shaped obstacle.

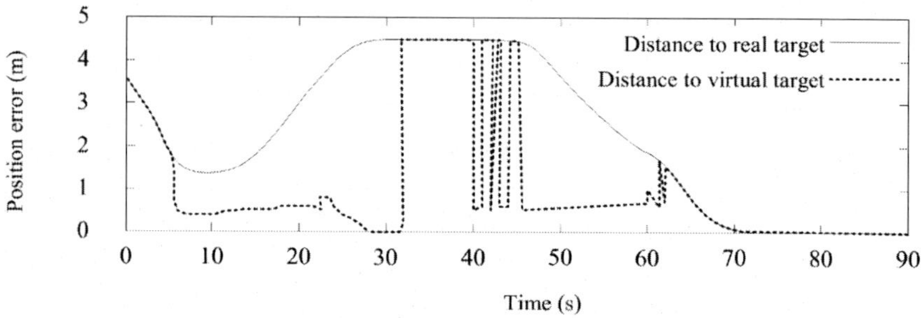

Figure 135. Time position error for simulation in a U-shaped obstacle.

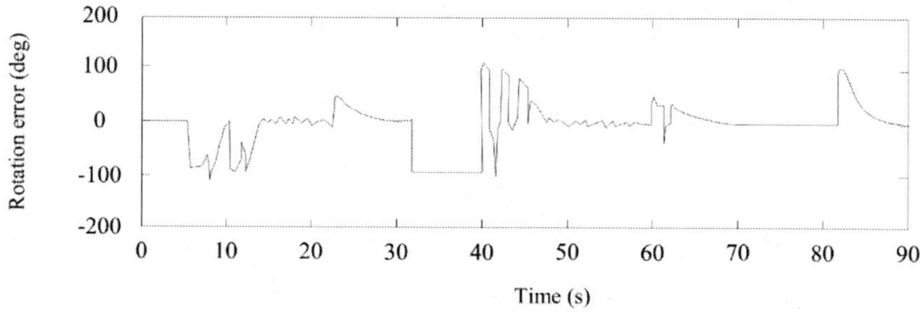

Figure 136. Rotation error for simulation in a U-shaped obstacle.

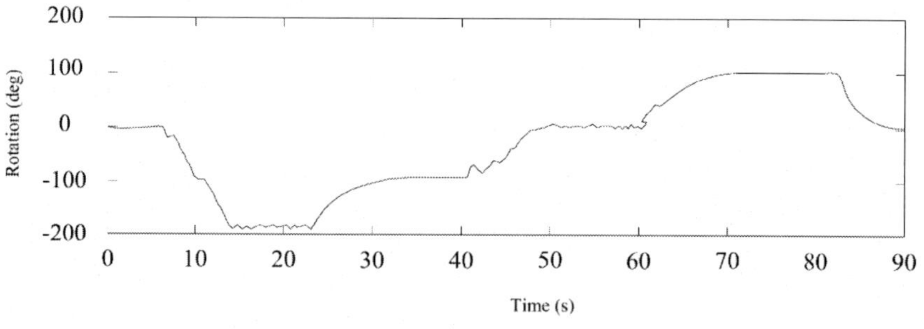

Figure 137. Rotation angle in terms of time for simulation in a U-shaped obstacle.

ROBOT PATH PLANNING USING VISION SENSORS

Three distinct sensors, namely occupancy sensor, distance sensor, and vision sensor, are utilized in sensor-based path-planning applications. Although occupancy sensor can extend the path, in-depth literature survey reveals that occupancy-related information is not technically demanding in a domain of information fusion since it might be either diagnostic or inadequate. For this reason, methods relying just on these sensors are far-fetched. The distance sensors receive the reliable body of data contrary to the occupancy sensor, but recognition of the danger zones for the robot through the extracted data could hardly be achieved. The vision sensor would be the optimal approach in this regard as a result of the useful information they provide.

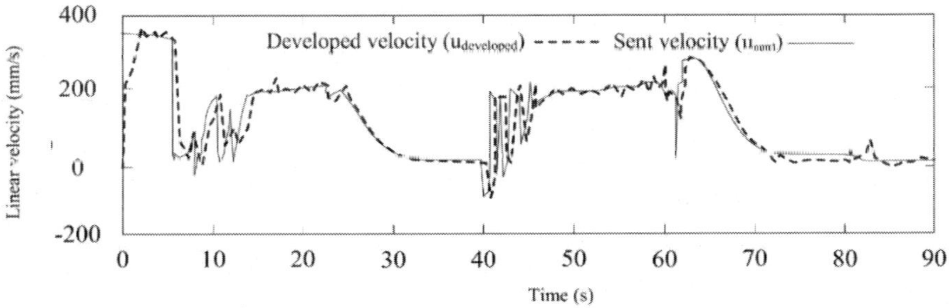

Figure 138. Linear velocity in terms of time for simulation in a U-shaped obstacle.

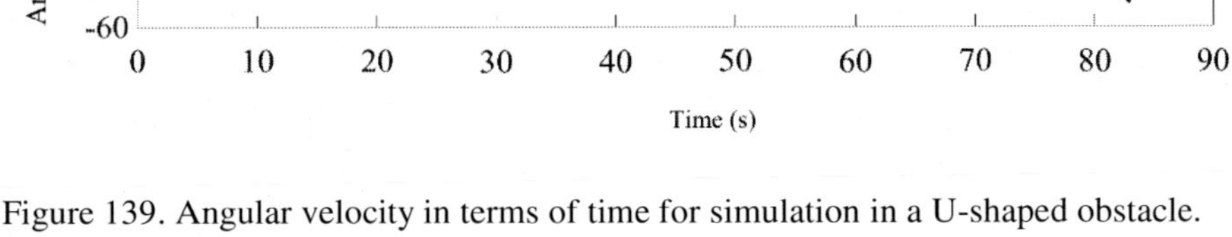

Figure 139. Angular velocity in terms of time for simulation in a U-shaped obstacle.

The Color Models

The ability to differentiate between colors is pertinent to the sufficient difference in the frequency and wavelength of any given color.

Consequently, quantification and storage of the frequency vector and the corresponding light intensity can be implemented in order to display pixels. Current technology using in real-time color sensors do not recognize minor differences within a full gamut of colors for image segmentation. As a result, the well-stablished methods to specify and store pixels have been developed. The RGB model is the most conventional color representation, in which each color is encoded by the intensity level of three additive primary colors, red, green, and blue. These primary colors could be added together in different combinations; a mixture of two pure additive primaries creates a subtractive color. Yellow, turquoise, and purple are three subtractive colors. The white color refers to the overlaying of three additive primary colors and the absence of all these three additive primary colors stands for the black color. Typically, the color intensity value of any pixel would be an unsigned integer number varying in the range of 0 to 255. To obtain light intensity (regardless of color), the mean intensity of the three additive primary colors should be measured. The human eye has light-sensitive photoreceptor cells, which respond to the frequency of red, green, and blue lights; therefore, the RGB model functions analogous to light perception in human beings.

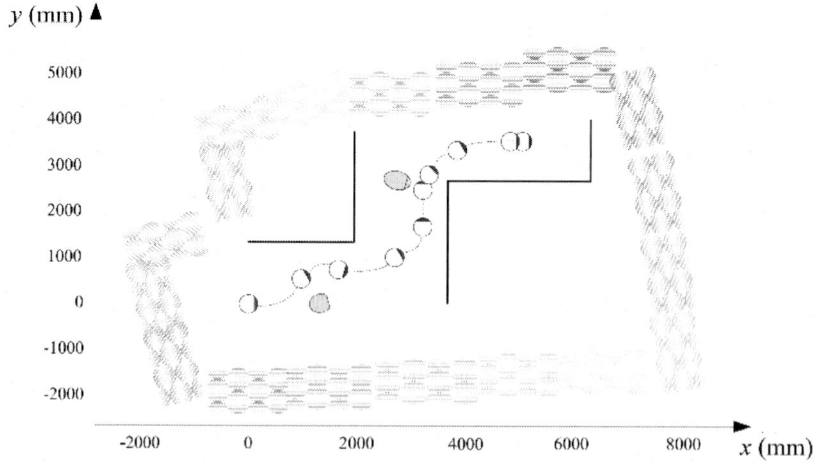

Figure 140. Tangential escape algorithm applied in an environment with a Z-shaped obstacle.

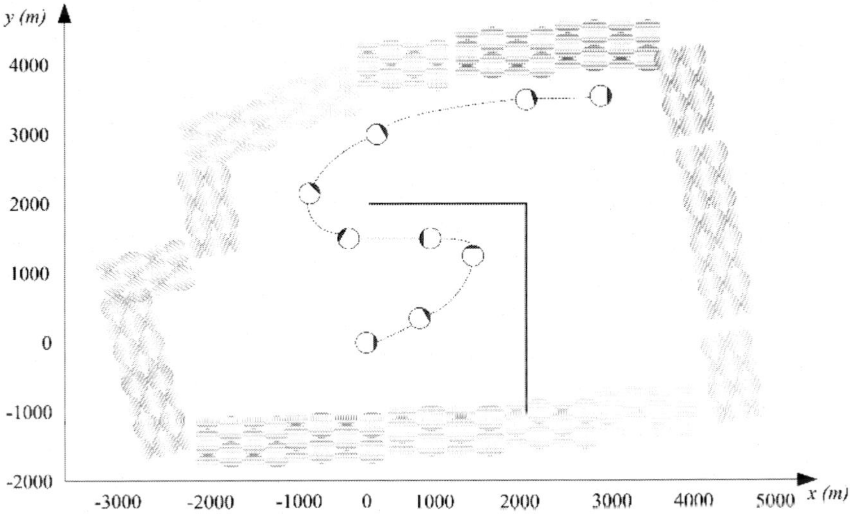

Figure 141. Tangential escape algorithm applied in an environment with a L-shaped obstacle.

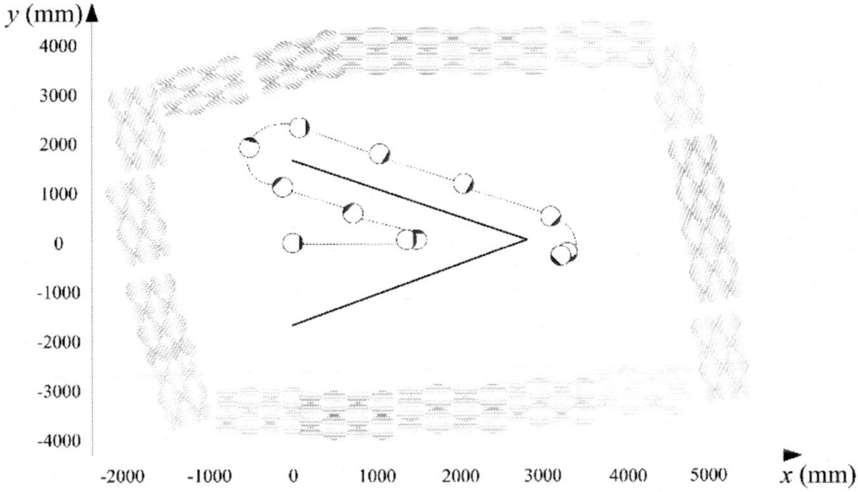

Figure 142. Tangential escape algorithm applied in an environment with a V-shaped obstacle.

Distinctive models depending on the kind of vision sensors applying in the design of robots must be adopted. In this case, there should be a tool to transform their used color model as a first step. For instance, the Nao humanoid robot takes advantages of the 422 YUV color model.

By applying Eq. (43), the 422 YUV color model can be transformed to the RGB.

$$\begin{bmatrix} R \\ G \\ B \end{bmatrix} = \begin{bmatrix} 1 & 0 & 1.13983 \\ 1 & -0.39465 & -0.58060 \\ 1 & 2.03211 & 0 \end{bmatrix} \begin{bmatrix} Y \\ U \\ V \end{bmatrix} \quad (43)$$

Low-Pass Filter

The noise in the sensors is considered normal due to the electromagnetic nature of light. Moreover, there is also an approximation continuity of color thanks to the continuously distributed of the solid bodies. Respectively, the presence of a high-frequency signal (the color variation of a pixel comparative to its neighbors) is likely to be noise that should be eliminated by designing a low-pass filter.

Segmentation and Mode Filter

According to the aforementioned paragraphs, each pixel is regarded as a combination of three intensity levels of red, green, and blue, which varies between 0 and 255. Even though these data are warranted to display the image, they do not play a prominent role in the path planning process, inasmuch as determination of robot's path is merely based on the type of pixel and has nothing to do with the detailed data. For the fulfillment of this aim, it must be accounted for whether each pixel connotes obstacle, danger, free, or target spaces. The conversion of the image from the color space to the surrounding environment is known as image segmentation. Various techniques to segment an image have been recommended; however, selecting the proper segmentation method highly depends on the position of the robot. All segmentation techniques inevitably contain an error (noise) which must be eliminated whilst passing the image through the Mode filter.

Expansion

In a time of the recording process, some parts of the external boundary of the obstacles may be stored in lieu of other spaces (mainly free space). Therefore, obstacles are slightly widened to eschew collisions. Furthermore, the expansion of the obstacles intensifies the effect of the miniature obstacles. Figure 143 demonstrates image expansion.

Schematic Structure of Vision System

The humanoid robot implemented in this project makes use of a camera to glean surrounding information. After capturing an image by the camera, it proceeds through a filter to obviate the attendant noise. Then in the ensuing steps, the segmentation of the image and passing it through a mode filter in order to remove the impact of noise generated during the segmentation are accomplished, consecutively. Eventually, the final image is produced utilizing the dilation process. The simplified structure of the vision system is depicted in Figure 144.

Path Planning in the Absence of Danger Space

In this section, four approaches, including synthetic potential field method, linguistic method, Markov decision processes, and fuzzy Markov decision processes, are evaluated and employed on the Aldebaran humanoid robot–Nao H25 V4.

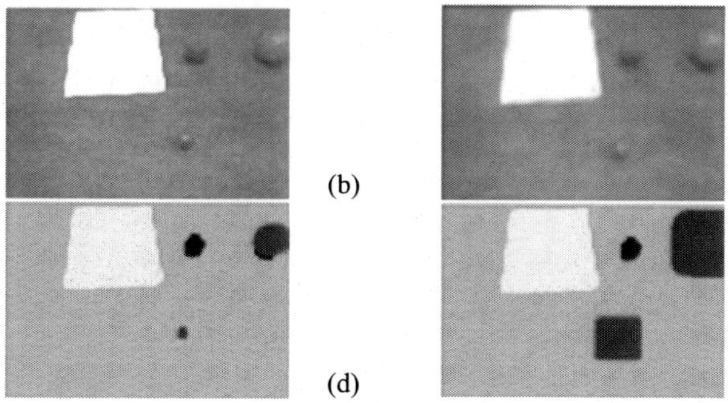

(a) (b)

(c) (d)

Figure 143. Image expansion to enhance the obstacle effect (a: main image; b: filtered image; c: segmented and filtered image; d: expanded image).

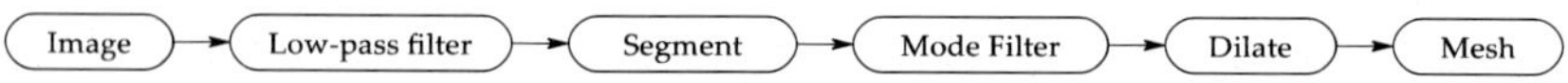

Figure 144. Schematic structure of the vision system.

Synthetic Potential Field Method

The electrostatic force exerted on two identical electrical particles is tantamount to Eq. (44):

$$\vec{F} = k\frac{q_1 q_2}{|r|^2}\overrightarrow{e_r} \tag{44}$$

in this equation, $\vec{F}$ is the exerted force, k is Columb constant, q_1 and q_2 are electrical charges of particles, the distance between them is r, and $\overrightarrow{e_r}$ is the unit vector connecting these electrical charges. It is postulated that the robot and obstacles have negative charges, and the target carries a positive one as well. Consequently, obstacles and the robot repulse each other, but the target attracts the negatively charged robot. On the ground of electrostatic laws, it is concluded that the accumulated forces of a negative charge on which one positive and N negatives charges exert, can simply calculated by Eq. (45):

$$\vec{F} = \overrightarrow{F_a} + \sum_{k=1}^{N}\overrightarrow{F_d}(k) \tag{45}$$

where F_a and F_d present instant attraction and repulsion forces, respectively. Secondly, the number of obstacles denote as k. Natural surrounding space is confined to each cell of meshed space, so Eq. (45) is reformulated as Eq. (46).

$$\vec{F} = \overrightarrow{F_a} + \sum_{k=1}^{N} \overrightarrow{F_d}(i,j) \tag{46}$$

in which (i,j) notes the respective cell's position. At that, n means the number of cells in a row, and $k = nj + i$. The vector decomposition along the global axes x and y is resulted in procuring the resultant forces F_a and F_d as below:

$$\overrightarrow{F_a} = F_{ax}\mathbf{i} + F_{ay}\mathbf{j} \tag{47}$$

$$\overrightarrow{F_d}(i,j) = F_{dx}(i,j)\mathbf{i} + F_{dy}(i,j)\mathbf{j} \tag{48}$$

The equivalent components thereupon are:

$$F_{ax} = \overrightarrow{F_a}.\frac{\vec{x}_{goal}}{|r_{goal}|} \tag{49}$$

$$F_{ay} = \overrightarrow{F_a}.\frac{\vec{y}_{goal}}{|r_{goal}|} \tag{50}$$

$$F_{dx}(i,j) = \overrightarrow{F_d}(i,j).\frac{\vec{x}(i,j)}{|r_{(i,j)}|} \tag{51}$$

$$F_{dy}(i,j) = \overrightarrow{F_d}(i,j).\frac{\vec{y}(i,j)}{|r_{(i,j)}|} \tag{52}$$

Substitution of Eqs. (49) and (50) for Eq. (47), and inserting Eqs. (51) and (52) into Eq. (48), Eqs. (53) and (54) can be readily procured as follows:

$$\overrightarrow{F_a} = \overrightarrow{F_a}.\frac{\vec{x}_{goal}}{|r_{goal}|}\mathbf{i} + \overrightarrow{F_a}.\frac{\vec{y}_{goal}}{|r_{goal}|}\mathbf{j} \tag{53}$$

$$\overrightarrow{F_d}(i,j) = \overrightarrow{F_d}(i,j).\frac{\vec{x}_{goal}}{|r_{goal}|}\mathbf{i} + \overrightarrow{F_d}(i,j).\frac{\vec{y}_{goal}}{|r_{goal}|}\mathbf{j} \tag{54}$$

The final product of considering Eq. (53) concurrently with Eq. (54) in Eq. (46) can be composed as follows:

$$\vec{F} = \overrightarrow{F_x} + \vec{F_y} \tag{55}$$

where

$$\overrightarrow{F_x} = \left(\overrightarrow{F_a}.\frac{\vec{x}_{goal}}{|r_{goal}|} + \sum_{k=1}^{N} \overrightarrow{F_d}(i,j)\,.\frac{\vec{x}(i,j)}{|r_{(i,j)}|}\right)\mathbf{i} \tag{56}$$

$$\overrightarrow{F_y} = \left(\overrightarrow{F_a}.\frac{\vec{y}_{goal}}{|r_{goal}|} + \sum_{k=1}^{N} \overrightarrow{F_d}(i,j)\,.\frac{\vec{y}(i,j)}{|r_{(i,j)}|}\right)\mathbf{j} \tag{57}$$

Eq. (67) is written down for obstacles concerning the fact that all obstacles are the same (attributed as obstacle space) as the following equation:

$$\overrightarrow{F_d}(i,j) = -k_d \frac{1}{(|r_{(i,j)}|)^2}\overrightarrow{e_r} \tag{58}$$

Considering the positively charged target point, it is feasible to rewrite the attraction equation as follows:

$$\overrightarrow{F_a} = k_a \frac{1}{(|r_{goal}|)^2}\vec{e}_r \tag{59}$$

Replacement of Eqs. (59) and (58) for Eq. (56) and Eq. (57), Eqs. (60) and (61) can be derived as follows:

$$\vec{F_x} = \left(k_a.\frac{\vec{x}_{goal}}{(|r_{goal}|)^3} - k_d \sum_{k=1}^{N} \frac{\vec{x}(i,j)}{(|r_{(i,j)}|)^3}\right)\mathbf{i} \tag{60}$$

$$\vec{F_y} = \left(k_a.\frac{\vec{y}_{goal}}{(|r_{goal}|)^3} - k_d \sum_{k=1}^{N} \frac{\vec{y}(i,j)}{(|r_{(i,j)}|)^3}\right)\mathbf{j} \tag{61}$$

Aside from the target point, any individual cell may mutually share obstacle and free space with a degree of uncertainty. So, the multiplication of magnitude of corresponding repulsive force and the membership function of obstacle space (μ) is used to fuzzify Eqs. (60) and (61) [72]. Thus, Eqs. (62) and (63) can be calculated as:

$$\vec{F_x} = \left(k_a.\frac{\vec{x}_{goal}}{(|r_{goal}|)^3} - \mu k_d \sum_{k=1}^{N} \frac{\vec{x}(i,j)}{(|r_{(i,j)}|)^3}\right)\mathbf{i} \tag{62}$$

$$\vec{F_y} = \left(k_a.\frac{\vec{y}_{goal}}{(|r_{goal}|)^3} - \mu k_d \sum_{k=1}^{N} \frac{\vec{y}(i,j)}{(|r_{(i,j)}|)^3}\right)\mathbf{j} \tag{63}$$

To meet the target's position, the robot requires to navigate its way through the direction of the force. For this reason, the robot angle is calculated by the following equation as specified:

$$\phi = \arctan2\left(\vec{F_y}, \vec{F_x}\right) \tag{64}$$

From the Eq. (64), "arctan2" is defined as Eq. (65):

$$\arctan2\left(\vec{F_y}, \vec{F_x}\right) = \begin{cases} \arctan\left(\frac{y}{x}\right) \; x > 0 \\ \arctan\left(\frac{y}{x}\right) + \pi \; y \geq 0; x < 0 \\ \arctan\left(\frac{y}{x}\right) - \pi \; y < 0; x < 0 \\ \frac{\pi}{2} \; y > 0; x = 0 \\ \frac{-\pi}{2} \; y < 0; x = 0 \\ \text{undefined} \; y = 0; x = 0 \end{cases} \tag{65}$$

The Rectifier

In the immediate vicinity of the target point, the attraction takes relatively large magnitudes, for the denominator of fractions in (62) and (63) tend to zero. Under this circumstance, it culminates in repulsive force from obstacles becoming highly ineffective with respect to the attraction force. In an attempt to curb this issue, it is propounded that the denominator of the related fraction is added to the integer 1.

$$\phi = \text{arctan2}(k_a.\frac{\vec{y}_{goal}}{(|r_{goal}|)^3+1} - \mu k_d \sum_{k=1}^{N} \frac{\vec{y}(i,j)}{(|r_{(i,j)}|)^3}, k_a.\frac{\vec{x}_{goal}}{(|r_{goal}|)^3+1}$$
$$-\mu k_d \sum_{k=1}^{N} \frac{\vec{x}(i,j)}{(|r_{(i,j)}|)^3}) \tag{66}$$

The result of this addition is illustrated in Figure 145.

Result of Synthetic Potential Field Method

As has been noted, Aldebaran humanoid robot–NAO H25 V4 is operated to figure out the validity of the suggested approach. Integration of themechanical structure of the employed humanoid robot such as implemented actuators and sensors are fully detailed [73]. By the same token, the study of the control mechanism of the robot is described [74]. Aldebaran's NaoQi as a benchmark and an extended code in C++ is used in order for the software architecture to develop. Correspondingly, Kubuntu 12.0.4 and Open CV 2.3.1 writing program in C++ in Qt creators is applied. In Figure 146, at the commencement of the process, the target point is regarded as a virtual one. The robot can by no means observe any obstacles in its path at the initial stage of simulation, that being so, the robot chooses a straight direction to reach the virtual target point. A distraction force will be added as soon as the robot encounters the first obstacle. The resultant force makes the robot pass across the gap between the two obstacles.

Linguistic Method

Having been laid a concrete foundation in the natural potential field, the linguistic methodologyserves as a tool to compute the field with linguistic rules rather than deterministic relations. The intensity of natural potential force and the synthetic potential field are proportional to the square of the distance and a descending function of distance, all at once. The image captured by the NAO's camera in conformity with the size of the NAO and the height of its camera is divided into 25 cells. Applying rules for the computation of the repulsive forces of the obstacles and the attractive force of the target point is briefly summarized in Tables 6 and 7 where Positive (P), Negative (N), Small (S), Zero (Z), Very (V), Medium (M) and Big (B) are the variables.

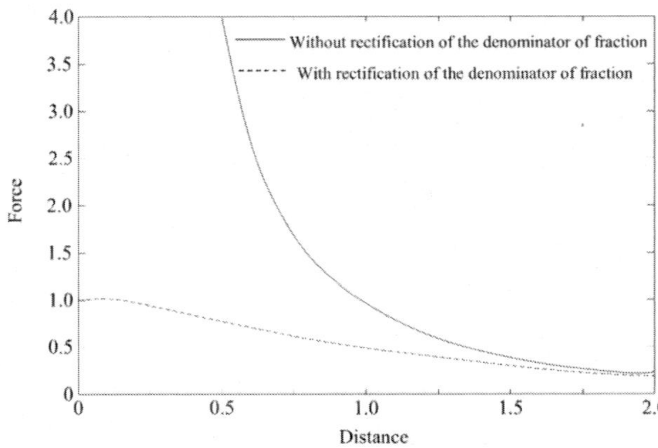

Figure 145. Attraction force in the presence and absence of the rectifier.

Not until the robot recognizes the target point through the fuzzification is a non-zero magnitude designated to the cells within which the target is located. This leads to the robot will become attracted to the target point. On the contrary, a virtual repulsive force exerted on the robot impedes its progress toward the target provided that the robot is unable to identify the target. In consequence, it is posited that a virtual target is situated in a cell which is in the closest proximity to the main target in order to effectively tackle the stated problem (Figure 147).

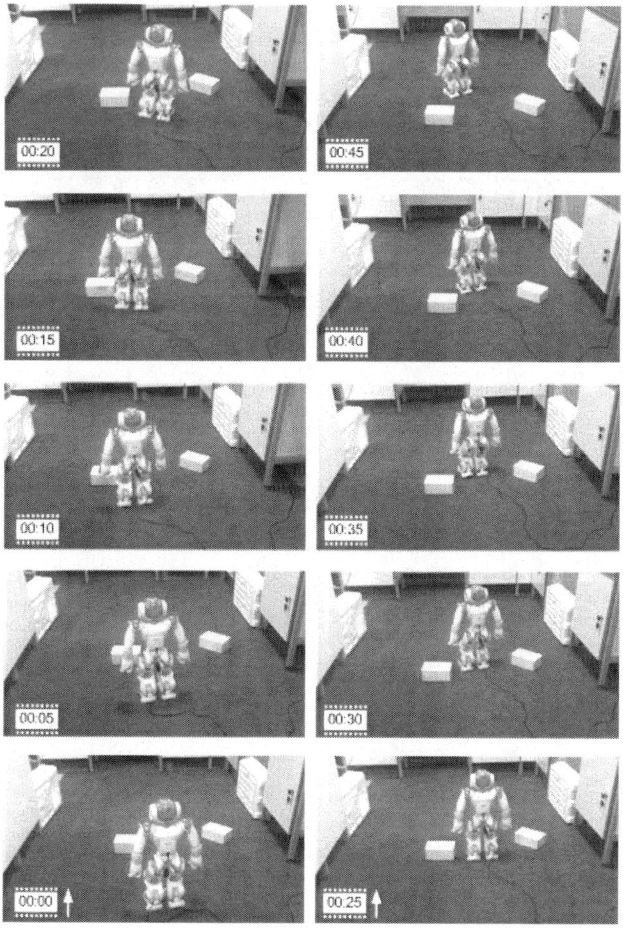

Figure 146. The traversed path by the Nao humanoid robot implementing the fuzzy synthetic potential field approach.

Table 6. The rules of obstacles' force in the direction of the *x*- and *y-axis*

j \ *i*	Axis *x*					Axis *y*				
	1	2	3	4	5	1	2	3	4	5
1	VSN	VSN	Z	VSP	VSP	VSP	VSP	VSN	VSN	VSN
2	VSN	VSN	Z	VSP	VSP	VSN	SN	SN	SN	VSN
3	SP	SN	Z	SP	SP	SN	MN	MN	MN	SN
4	MN	MN	Z	MP	MP	MN	BN	VBN	BN	MN
5	BN	VBN	Z	VBP	BP	MN	VBN	VBN	VBN	MN

Table 7. The rules of the target's force in the direction of the *x*- and *y*-axis

j \ *i*	Axis *x*					Axis *y*				
	1	2	3	4	5	1	2	3	4	5
1	VSP	VSP	Z	VSN	VSN	VSP	VSP	VSP	VSP	VSP
2	VSP	VSP	Z	VSN	VSN	VSP	SP	SP	SP	VSP
3	SP	SP	Z	SN	SN	SP	MP	MP	MP	SP
4	MP	MP	Z	MN	MN	MP	BP	VBP	BP	MP
5	BP	VBP	Z	VBN	BN	MP	VBP	VBP	VBP	BP

With regard to n and m, the total number of fuzzy rules determining the level of the output is 4mn. In the natural potential field, the net force acted on a single charged particle is the sum of forces which would have been generated by other charged particles individually (superposition principle). That is to say; this axiom can be applied to the synthetic potential field approach. In this method, the superposition force is commensurate with the weighted average defuzzification that can be explicated as below:

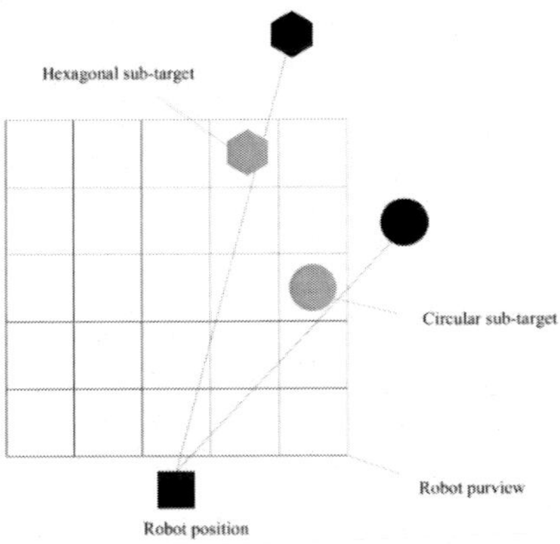

Figure 147. The robot's position and sub-targets in the purview of the robot.

$$f_x = \frac{\sum_{i=1}^{N} P^i_{\text{obstacle}} \hat{f}^i_{rx} + P^i_{\text{target}} \hat{f}^i_{ax}}{\sum_{i=1}^{N} P^i_{\text{obstacle}} + P^i_{\text{target}}} \tag{67}$$

$$f_y = \frac{\sum_{i=1}^{N} P^i_{\text{obstacle}} \hat{f}^i_{ry} + P^i_{\text{target}} \hat{f}^i_{ay}}{\sum_{i=1}^{N} P^i_{\text{obstacle}} + P^i_{\text{target}}} \tag{68}$$

where P^i_{obstacle} denotes the obstacle presence probability, P^i_{target} refers to the target presence probability. Then f_x and f_y represent, respectively, the force components along the x- and y-axis.

Simplification

In the proposed method, the robot moves along the direction of the field. Presumably, knowing the right direction of the field per se is sufficient. It should be noticed that the denominators of fractions are alike in Eqs. (67) and (68), by taking advantage of this fact, both equations are directly multiplied into the common denominator. In this case, without changing directions of the forces, equations can be simplified considerably as:

$$f'_x = \sum_{i=1}^{N} P^i_{\text{obstacle}} \hat{f}^i_{rx} + P^i_{\text{target}} \hat{f}^i_{ax} \tag{69}$$

$$f'_y = \sum_{i=1}^{N} P^i_{\text{obstacle}} \hat{f}^i_{ry} + P^i_{\text{target}} \hat{f}^i_{ay} \tag{70}$$

The direction of the force can be found out as follows:

$$\phi = \text{arctan2}(f'_y, f'_x) \tag{71}$$

Result of Linguistic Method

Figure 148 displays the path traversed by the Nao robot based on the linguistic method. At the very beginning of the undertaken process, the robot is not able to see the target point. So, the robot goes for a sub-target in lieu of the target, and therefore, it seeks to approach the proximate target position. It is leading the robot moves within the separation distance between obstacles and then reaches the sub-target. After identifying the target by the robot, it moves to the target point without either striking obstacles or deviation.

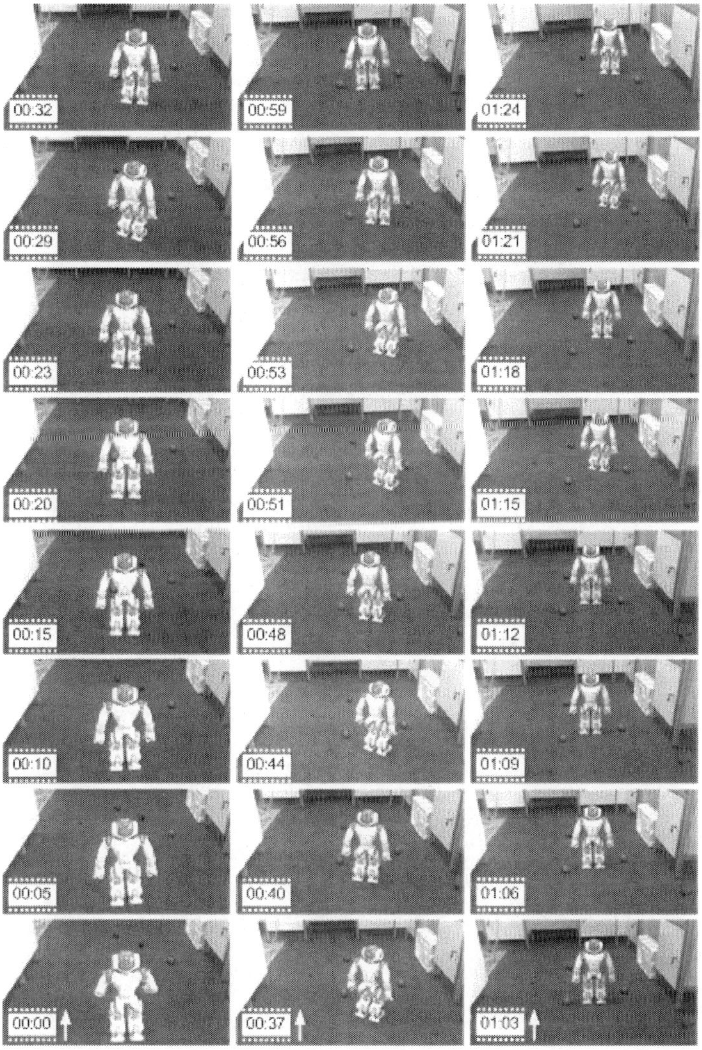

Figure 148. The path traversed by the Nao humanoid robot utilizing the linguistic approach.

Markov Decision Processes

A generalization of Markov chains is the Markov decision processes that laid a mathematical cornerstone for modeling decision making in conditions in which the outcomes are partly random and out of control [75]. To put it differently, the Markov decision process is a discrete-time stochastic control process. Thereby, in every time step, the decision maker selects the best course of action among the probable courses of actions regarding in terms of the state of the process defining as s. Henceforth, The process performs by randomly moving to a new state s', and giving an equivalent reward R to the decision maker [76]. On that account, the probability of making the transition to a certain state of the process is a function of the chosen action. This implies that the state s' depends on not only the state s, but also the action of the decision-maker a. Meanwhile, a and s are equally independent of all past actions and states. To put it briefly, proceeding from one state to another in Markov decision exhibits the Markov property [77-79]. The major problem of Markov decision processes is to determine a function π with which the decision maker is able to choose the action $\pi(s)$ in state s. Indeed, the prime objective is to find out a policy π maximizing the cumulative function of the random rewards. Besides, the value function evaluates the magnitude of expected future rewards, which a system gets only if it works from state s and follows the policy. Given this, each policy contributes to a value function as below:

$$V^{\pi}(s_0) = E[R(s_0) + \gamma R(s_1) + \gamma^2 R(s_2) + \cdots |\pi]$$
$$= E[\textstyle\sum_{t=0}^{N} \gamma^t R(s_t)|\pi] \tag{72}$$

This equation can be revised as:

$$V^{\pi}(s_0) = E[R(s_0) + \gamma(R(s_1) + \gamma R(s_2) + \cdots |\pi] \tag{73}$$

Equation (73) is named for Bellman and is reduced to:

$$V^{\pi}(s) = E[R(s) + \gamma \textstyle\sum_{s'} P(s, a, s')V^{\pi}(s')] \tag{74}$$

The optimum policy π^* and the optimum value function V^* are respectively presented in Eqs. (75) and (76) as:

$$V^*(s) = R(s) + max_a \gamma \sum_{s'} P(s,a,s') V^*(s') \quad (75)$$

$$\pi^*(s) = argmax_a \sum_{s'} P(s,a,s') V^*(s') \quad (76)$$

Assumed that, a system with a system with n states lead n equations which constitute a solvable system of equations.

Path Planning

The direction of movement of the robot in any given step employing the Markov decision processes can be feasibly determined. To do so, a negative reward will bc given if the robot collides with an obstacle. On the other hand, to reach the target point augurs a positive reward. Having this in mind, the robot may tend to stay still in some episodes for the sakc of not gaining the negative reward especially adjacent to an obstacle. In response to this concern, the free space to which a small negative reward is attributed when compared to the significant negative reward of obstacles.

The complex computation of the reward depends on the observation of the target point. Should the robot see the target, it will count on the provided information from the image. But if despite the obtained information from the image, the robot cannot observe the target point, it requires both its own and the target's coordinates to create a sub-target.

A sub-target, which might be defined as a virtual target, could steer the robot toward the tangible target. Figure 149 reveals the calculation of the state for the sub-targets. As reflected in this figure, the black shapes like the black hexagon are defined as the targets. In the same vein, the gray shapes, for instance, the gray circle, are selected for the sub-targets.

Not until the robot can see the obstacle in each situation do the free space, the obstacle, and the sub-target have $-w_1$, $-w_2$ and +1 point, respectively. This being the case, the reward function could be measured as follows:

$$R(i,j) = P_{sub-target}(i,j) + w_1 P_{freespace}(i,j) + w_2 P_{obstacle}(i,j) \quad (77)$$

On the other side, in situations where the robot can see the obstacle, the free space has $-w_1$ point, the obstacle has $-w_2$ point, and the target has $+1$. So, the reward function could be calculated as follows:

$$R(i,j) = P_{target}(i,j) + w_1 P_{freespace}(i,j) + w_2 P_{obstacle}(i,j) \quad (78)$$

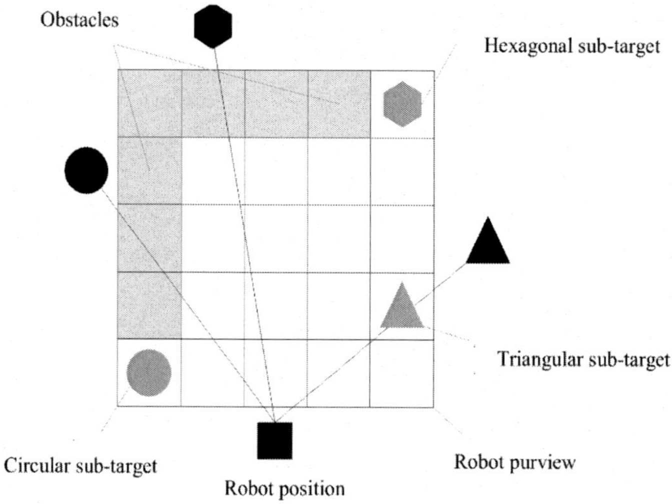

Figure 149. The appointing sub-targets for the main target procedure.

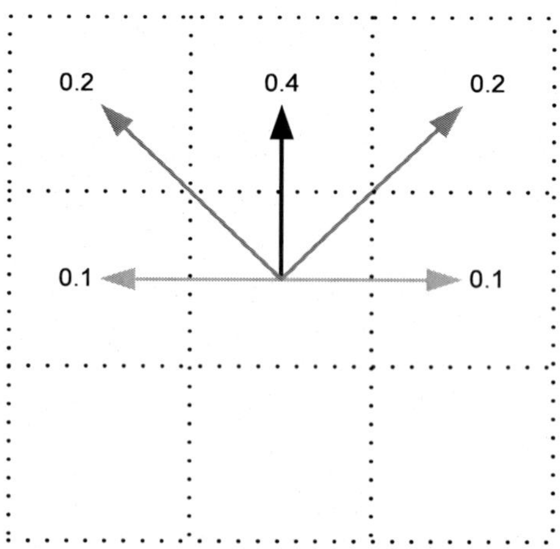

Figure 150. The prospect of moving to other states apropos of selecting forward movement.

In a number of situations, as in Figure 149, the target is outside of the robot sight range. Hence, a sub-target is utilized on the main target's behalf. Figure 150 vividly depicts the likely movement function of the robot, speculating that the robot keeps moving forward.

The Bellman equation (Eq. (75)) is nonlinear and difficult to resolve the problem. In this instance, with the purpose of procurement of the optimal value function, the next algorithm is practiced instead of directly solving the Bellman equation:

```
Input: Reward function R(s)
      Output: Value function V(s)
      begin
        ∀s V(s) → 0
        repeat
          for all the s do
                R(s) + max_a γ∑P(s,a,s')V(s') → B(s)
          end
        B(s) → V(s)
          until V(s) converges;
      end
```

Markov decision processes, invariably, offer an optimal path rested on the current state while the optimal path is not perpetually the best choice; since the robot does not ever enjoy a panoramic view of its environment. As an illustration, when the robot progresses along a wall and is interrupted to see the target, the robot paradoxically struggles with the intention of approaching directly to the target. In this situation, a grave collision may occur between the robot and the wall. A rectifier can readily decipher this issue by warning the robot about lateral obstacles and so avoiding a collision in its next steps.

Results of Markov Decision Processes

Figures 151 and 152 portray the final results of the application of the Markov decision processes to the Nao humanoid robot. In these two figures, obstacles have different configurations. As evidenced by these figures, the robot satisfactory approaches the target without colliding with obstacles.

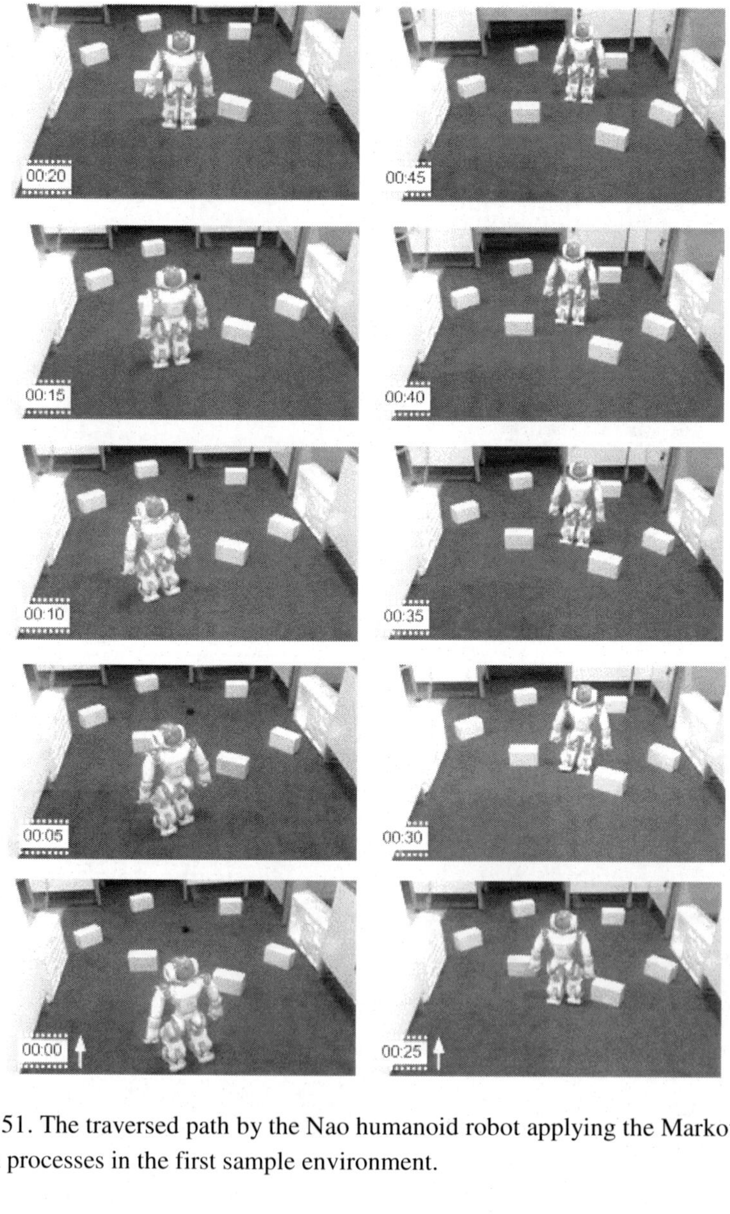

Figure 151. The traversed path by the Nao humanoid robot applying the Markov decision processes in the first sample environment.

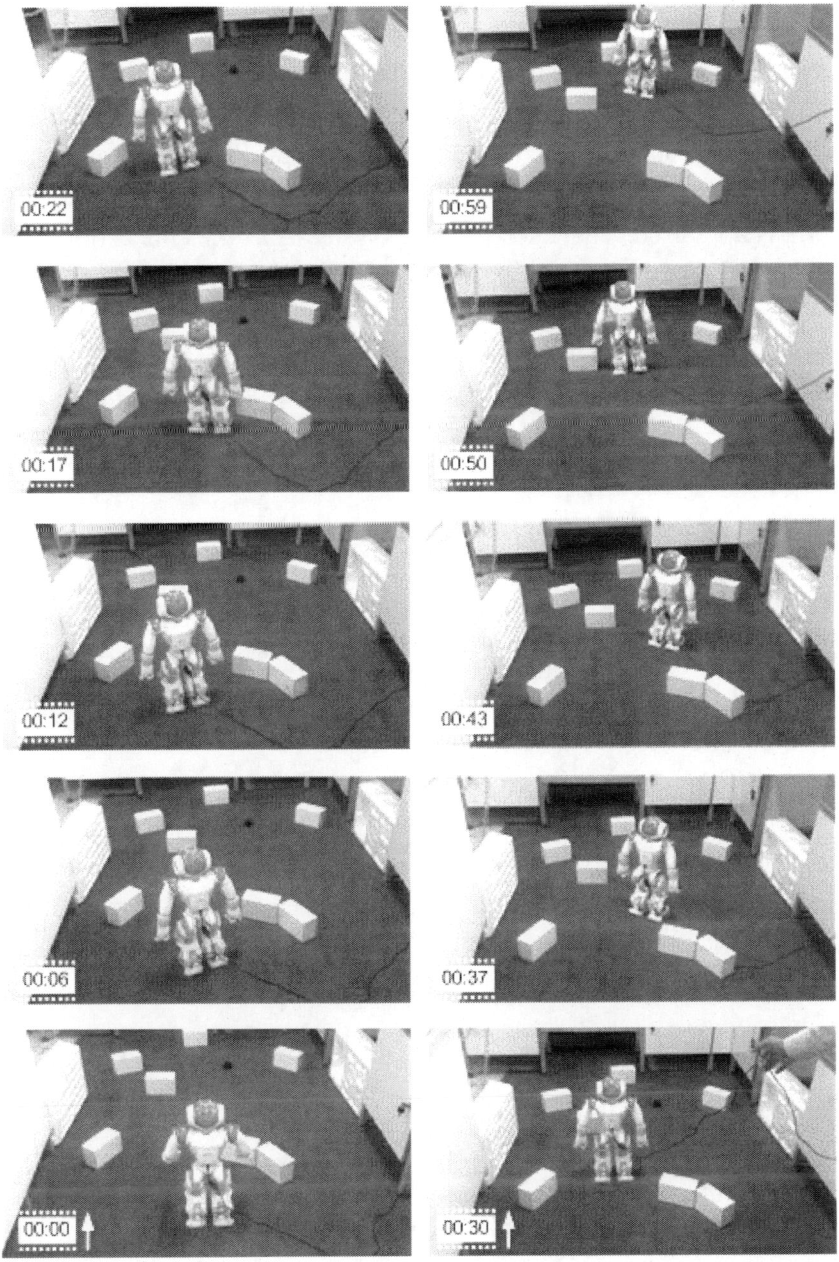

Figure 152. The path traversed by the Nao humanoid robot applying the Markov decision processes in the second sample environment.

Fuzzy Markov Decision Processes

As has already been pointed out, the policy of the Markov decision processes is premised upon the choice of an action through which the highest reward can be achieved. Here the classic decision-making framework is being adjusted. In so doing, in the first place, the value function is appraised. Thereafter, a fuzzy system specifies the action based upon the function. In accordance with the reward, the value function used in separate steps takes different intervals. In the fuzzy inference method, the crisp inputs through fuzzification are converted into the fuzzy membership value between 0 to 1. In light of this, the cost function must be normalized first of all. Ergo, the normalized value function can be gained using the following equation:

$$V^{new}(i,j) = \frac{V^*(i,j)-b}{a-b};\ a = \max_{(i,j)} V^*(i,j);\ b = \min_{(i,j)} V^*(i,j) \quad (79)$$

where $V^*(i,j)$ indicates the value function in each step, a and b are the maximum and the minimum of $V^*(i,j)$, respectively. As a rule, the inputs of the fuzzy inference method are an immediate neighbor of the position of the robot. In the classical approach, only the closest robot neighbors are selected as optimal choices. however, the fuzzy Markov decision processes stand in marked contrast to this view, where the neighboring radius is wider. The square of the Euclidean distance between each cell and the position of the robot is summarized in Table 8. The modification of Table 8 is provided in Fig. 153 in which four neighborhoods with different radii are exemplified.

In this respect, the neighborhood radius is supposed to be $\sqrt{10}$, and then after that, the inputs of the fuzzy inference method are produced using Table 9.

Subsequently, the new value function, which may be taken into consideration as the membership function, is fuzzified. This can activate over-fuzzification phenomenon for the path, which creates unwelcome complexity. To overcome this handicap, other fuzzification functions (e.g., exponential function) could be characterized. In particular, it is quite viable to express the membership function like this:

Table 8. Square of distance between each cell and the robot's position

	1	2	3	4	5
1	29	26	25	26	29
2	20	17	14	17	20
3	14	10	9	10	14
4	8	5	4	5	8
5	5	2	1	2	5
			Robot		

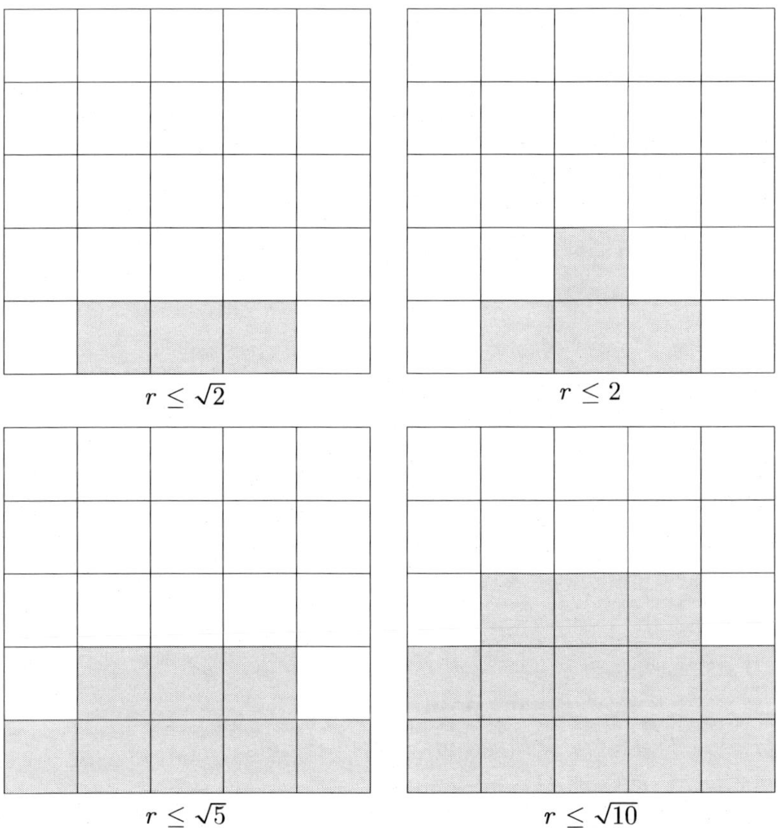

Figure 153. Four different neighborhoods of the robot in its frontal view.

Table 9. Inputs of the fuzzy inference method with a neighborhood radius of $\sqrt{10}$ for a robot with 25 cells of view

	1	2	3	4	5
1					
2					
3		X_1	X_2	X_3	
4	X_4	X_5	X_6	X_7	X_8
5	X_9	X_{10}	X_{11}	X_{12}	X_{13}
			Robot		

$$\mu(A_i) = x_i^n \tag{80}$$

in which x_i represents the probability of x in cell i and n is an integer number. Selection of the correct integer necessitates a wealth of experiment and pertains to the environment, plus the robot's camera. Each cell is governed by a fuzzy rule. The rule calculates the angle ϕ, which finds the direction of the robot's motion. The right-hand angles and the left-hand angles have positive and negative degree measure, respectively. Moreover, the head-on direction is zero all the time.

1. If A_1=1, then ϕ is a very small positive angle.
2. If A_2=1, then ϕ is zero.
3. If A_3=1, then ϕ is a very small negative angle.
4. If A_4=1, then ϕ is a medium positive angle.
5. If A_5=1, then ϕ is a small positive angle.
6. If A_6=1, then ϕ is a zero angle.
7. If A_7=1, then ϕ is a small negative angle.
8. If A_8=1, then ϕ is a medium negative angle.
9. If A_9=1, then ϕ is a big positive angle.
10. If A_{10}=1, then ϕ is a medium positive angle.
11. If A_{11}=1, then ϕ is a zero angle.
12. If A_{12}=1, then ϕ is a medium negative angle.
13. If A_{13}=1, then ϕ is a big negative angle.

With an eye toward several current defuzzification methods, the weighted average is carefully set forth and described hereafter.

$$\phi = \frac{\sum \mu(A_i)\widehat{\phi_i}}{\sum \mu(A_i)} \tag{81}$$

where $\mu(A_i)$ refers to the membership function and $\widehat{\phi_i}$ is the direction of the robot's motion.

Result of Fuzzy Markov Decision Processes in the Absence of Danger Space

The result of the implementation of the fuzzy Markov decision processes on the Aldebaran humanoid robot–Nao H25 is photographically reported in Figure 154. As shown in this figure, the robot successfully goes across obstacles and reaches the target.

PATH PLANNING IN THE PRESENCE OF DANGER SPACE

Generally speaking, a robot experiences three main kinds of environments: obstacle-free environment, obstacle, and target. Despite that, it is possible for the robot to encounter an additional type of environment intermittently. As an example, in a tree-covered area, trees may be deemed as obstacles, whereas mud pits do not restrict the robot's movement and the robot travels through it. Admittedly, it is not sensible to be on the verge of a muddy path in the accompany of the dry and flat ground. In such a case, mud pits are not permitted to be contemplated as the free spaces (as dry ground) nor are obstacles (as trees) allowed. Intending to unravel this problem, a new space titled danger space is inaugurated and supplemented with the three established environments. Danger space has the extensibility of encompassing other cases, notably the greasy floor in a factory and territories exposed to the enemy in a field of battle, to name but a few.

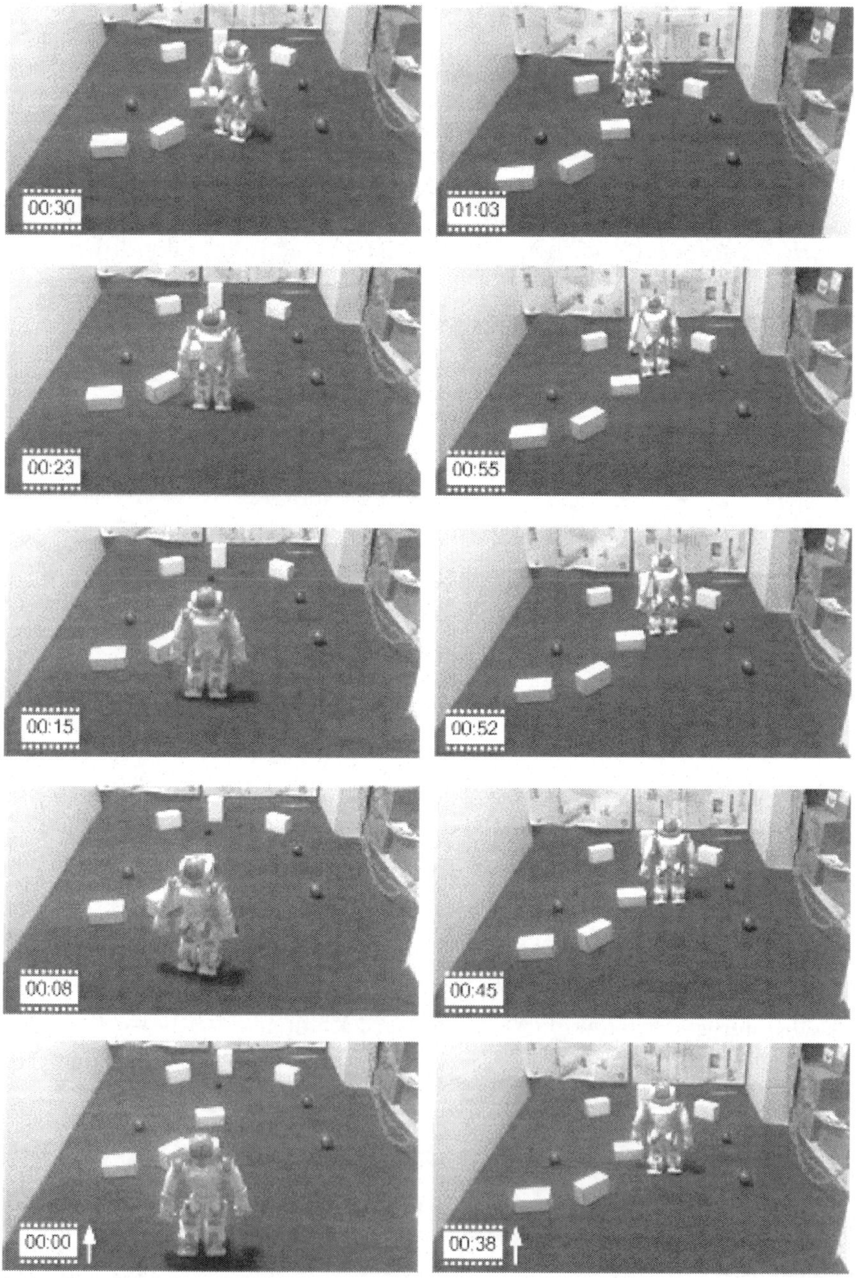

Figure 154. The path traversed by the Nao humanoid robot using the fuzzy Markov decision processes.

Disadvantages of Reward Calculation by Linear Relations

In the preceding section, Eq. (78) was advanced toward the calculation of rewards in cells excluding a danger area. Due to the linearity of the equations, this equation can be expanded to the danger space as follows:

$$R(i,j) = P_{goal}(i,j) + w_1 P_{freespace}(i,j) + w_2 P_{danger}(i,j) + w_3 P_{obstacle}(i,j) \quad (82)$$

This equation seems to be persuasive; its linear characteristic may nevertheless get the robot embroiled in difficulty. In order to exemplify this situation, let consider the coefficients of danger space and free space are almost close to one another, it appears to be irreproachable to move across the danger space to go through the free space, which is not a well-defined and apt choice. Another dilemma would be when the danger space whose coefficients are just as same as the space comprising the obstacle, is the only accessible option for the passage of the robot which is inclined to traverse the obstacle and finally collide the obstacle. As a consequence, an intelligent arrangement to determine rewards is a must.

Reward Calculation by the Fuzzy Inference System

At this very spot, the Gaussian membership functions are signified in the fuzzification process. Outlined seen in Figure 155, space approximately nearby a cell is a member of the fuzzy set consisting of zero, small, medium, big, very big.

The Takagi-Sugeno method is implemented to compute rewards applying the rules presented in Table 10. As a case in point, a literal interpretation of rule 1 envisages that if the prospect of the existence of the obstacle is between 0.375 and 1. In addition, the probability of the existence of the danger is between 0 and 0.375, the likelihood of the existence of the free space is likewise between 0 and 0.375, and the probability of the presence of the target is between 0 and 0.625, then the reward will be considered as o.1 Target.

The weighted average method for defuzzification of reward is defined as:

$$\text{Reward} = \frac{\sum \mu_i \hat{R}_i}{\sum \mu_i} \tag{83}$$

where "Reward" refers to the defuzzification of reward. The membership function and the reward of each fuzzy rule are μ_i and $\hat{R}_i$, respectively.

Schematic Structure of Fuzzy Markov Decision Processes

When the final image is produced, the calculation of the rewards germane to the individual cells of the image is initiated. Ever since then, Markov decision processes act as an input for the fuzzy inference system. The output of the fuzzy inference system equips the robot with the requisite information on which deciding about the direction and the robot's movement is based. The diagram structure of this process is encapsulated in Figure 156.

Results of Fuzzy Markov Decision Processes in the Presence of Danger Space

The practical outcomes of applying the fuzzy Markov decision processes on the Nao humanoid robot are demonstrated in Figures 157 and 158. According to Figure 157, at the start of the path, the robot needs to undergo the danger space as the solely available alternative, doing so necessarily entails being faced with obstacles and danger space. Only in this way can the robot pursue moving toward the target point. During the robot approaching the next danger space and due consideration of the availability of free space, it attempts to eschew the danger space and moving toward the target point. Apart from that, the robot makes an effort to choose the optimal path to reach the target. At this precise moment, the robot retraces a step to the right side and, conceding that it has accomplished extending through the danger space, turns left and walks towards the target. It induces the robot to pass the danger space during turning to come near the target. This problem can be rectified by enlarging the danger space. The robot bypassing the danger space, and successfully reaching the target is shown in Figure 158.

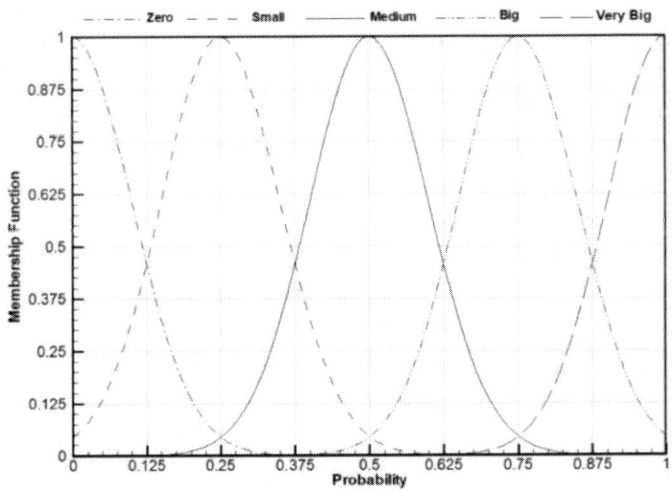

Figure 155. Fuzzy sets and membership functions of each space based on its probability.

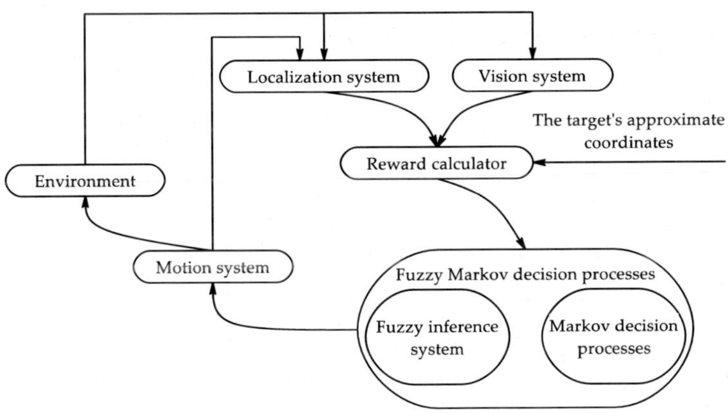

Figure 156. Schematic structure of fuzzy Markov decision processes.

Table 10. Fuzzy inference rule for reward

Rule	Obstacle	Danger	Free	Target	Reward
1	Big	Zero	Zero	Small	0.1Target
2	Medium			Medium	
3			Small	Small	
4		Small	Zero		
5	Small	Medium			
6		Zero	Medium		
7		Small	Zero	Medium	0.2Target
8		Zero	Small		
9		Zero	Zero	Big	0.25Target
10	Zero	Big	Zero	Small	0.333Target
11		Medium	Small	Small	0.5Target
12		Small	Medium	Small	0.667Target
13		Medium	Zero	Medium	0.75Target
14		Small	Small		
15		Zero	Big	Small	
16		Zero	Zero	Very Big	Target
17		Small		Big	
18		Zero	Medium	Medium	
19			Small	Big	
20	Small	Big	Zero	Zero	0.4Obstacle
21		Medium	Small		
22		Small	Medium		
23	Medium	Small	Small	Zero	0.6Obstacle
24	Big	Zero	Small	Zero	0.8Obstacle
25	Medium		Medium		
26	Very Big	Zero	Zero	Zero	Obstacle
27	Big	Small			
28	Medium	Medium			
29	Zero	Small	Big	Zero	0.25Danger
30		Medium	Medium	Zero	0.5Danger
31		Very Big	Zero	Zero	Danger
32	Small	Zero	Big		
33	Zero	Big	Small		
34		Zero	Very Big	Zero	Free

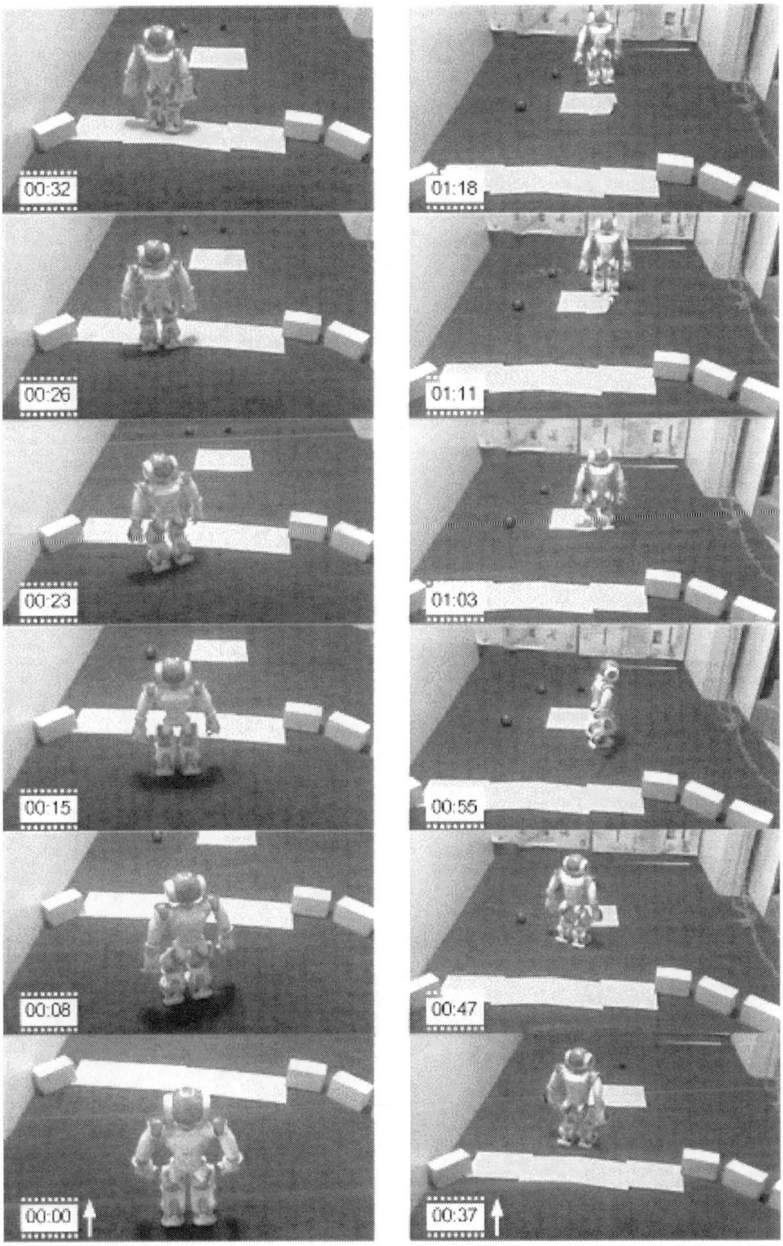

Figure 157. The path traversed by the Nao humanoid robot in presence of danger space using the fuzzy Markov decision processes in the first sample environment.

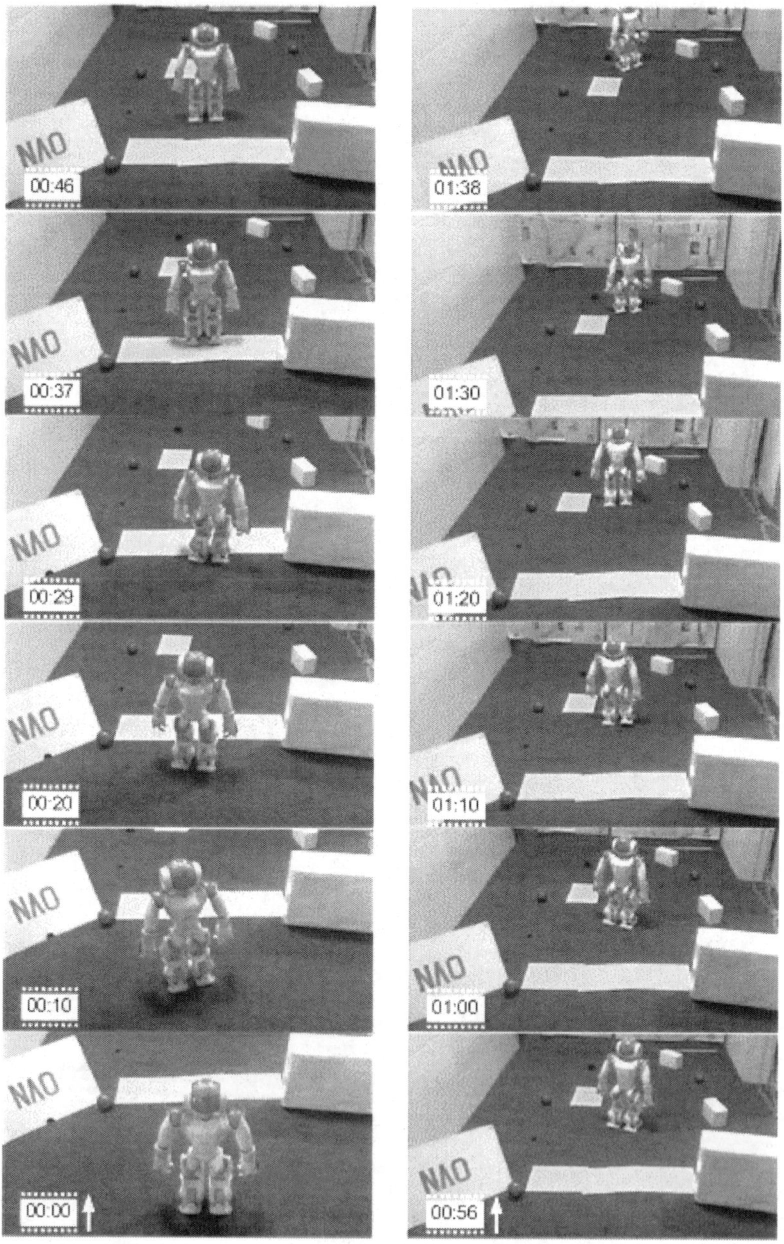

Figure 158. The path traversed by the Nao humanoid robot in presence of danger space using the fuzzy Markov decision processes in the second sample environment.

In the final analysis, the adoption of the Markov decision processes precipitates swifter performance in comparison with the other suggested methods. Equally important, the use of the fuzzy inference system brings about a smoother optimal path in relation to previous ones. Had further, the fuzzy Markov decision processes facilitates designing a path needless of impeccable information about all physical facets of the obstacle including its shape, position, and orientation, and of taking up the massive volume of memory conducive to record data acquisition from two-dimensional or three-dimensional maps as well.

REFERENCES

[1] Ng JS. *An analysis of mobile robot navigation algorithms in unknown environments*. 2010.

[2] Borenstein J, Koren Y. The vector field histogram-fast obstacle avoidance for mobile robots. *IEEE transactions on robotics and automation*. 1991;7:278-88.

[3] Bianco G, Cassinis R. Multi-strategic approach for robot path planning. *IEEE*. p. 108-15.

[4] Lumelsky V, Skewis T. A paradigm for incorporating vision in the robot navigation function. *IEEE*. p. 734-9.

[5] Lumelsky VJ. On the connection between maze-searching and robot motion planning algorithms. *IEEE*. p. 2270-5.

[6] Lumelsky V, Sun K. Gross motion planning for a simple 3D articulated robot arm moving amidst unknown arbitrarily shaped obstacles. *IEEE*. p. 1929-34.

[7] Sun K, Lumelsky V. Computer simulation of sensor-based robot collision avoidance in an unknown environment. *Robotica*. 1987;5:291-302.

[8] Lumelsky VJ. Effect of robot kinematics on motion planning in unknown environment. *IEEE*. p. 338-43.

[9] Lumelsky VJ. A comparative study on the path length performance of maze-searching and robot motion planning algorithms. *IEEE transactions on robotics and automation*. 1991;7:57-66.

[10] Skewis T, Lumelsky V. Experiments with a mobile robot operating in a cluttered unknown environment. *Journal of Robotic Systems*. 1994;11:281-300.

[11] Lucas C, Lumelsky V, Stepanov A. Comments on" Dynamic path planning for a mobile automation with limited information on the environment"[with reply]. *IEEE Transactions on Automatic Control*. 1988;33:511-2.

[12] Lumelsky VJ, Skewis T. Incorporating range sensing in the robot navigation function. *IEEE Transactions on Systems, Man, and Cybernetics*. 1990;20:1058-69.

[13] Lumelsky V, Tiwari S. An algorithm for maze searching with azimuth input. *IEEE*. p. 111-6.

[14] Lumelsky VJ. Dynamic path planning for a planar articulated robot arm moving amidst unknown obstacles. *Automatica*. 1987;23:551-70.

[15] Lumelsky VJ. *Continuous robot motion planning in unknown environment*. Adaptive and Learning Systems: Springer; 1986. p. 339-58.

[16] Lumelsky V, Stepanov A. Dynamic path planning for a mobile automaton with limited information on the environment. *IEEE Transactions on Automatic Control*. 1986;31:1058-63.

[17] Lumelsky VJ, Stepanov AA. Effect of uncertainty on continuous path planning for an autonomous vehicle. *IEEE*. p. 1616-21.

[18] Lumelsky VJ, Stepanov AA. Path-planning strategies for a point mobile automaton moving amidst unknown obstacles of arbitrary shape. *Algorithmica*. 1987;2:403-30.

[19] Abelson H, DiSessa AA. *Turtle geometry: The computer as a medium for exploring mathematics*: MIT press; 1986.

[20] Berge C. The theory of graphs: Courier Corporation; 2001.

[21] Lumelsky V. Algorithmic issues of sensor-based robot motion planning. *IEEE*. p. 1796-801.

[22] Lumelsky V. Continuous motion planning in unknown environment for a 3D cartesian robot arm. *IEEE*. p. 1050-5.

[23] Miskon MF, Russell AR. *Close Range Inspection Using Novelty Detection Results*. Springer. p. 947-56.

[24] Kamon I, Rimon E, Rivlin E. Tangentbug: A range-sensor-based navigation algorithm. *The International Journal of Robotics Research*. 1998;17:934-53.

[25] Kim S-K, Russell JS, Koo K-J. Construction robot path-planning for earthwork operations. *Journal of computing in civil engineering*. 2003;17:97-104.

[26] Pearl J. *Heuristics: intelligent search strategies for computer problem solving*. 1984.

[27] Nilsson NJ. *A mobile automaton: An application of artificial intelligence techniques*. SRI International Menlo Park CA Artificial Intelligence Center; 1969.

[28] Ko WS, Senevieatne LD, Earles SWE. Space representation and map building-A triangulation model to path planning with obstacle avoidance. *IEEE*. p. 2222-7.

[29] Foux G, Heymann M, Bruckstein A. Two-dimensional robot navigation among unknown stationary polygonal obstacles. *IEEE transactions on robotics and automation*. 1993;9:96-102.

[30] Choset HM, Hutchinson S, Lynch KM, Kantor G, Burgard W, Kavraki LE, et al. *Principles of robot motion: theory, algorithms, and implementation*: MIT press; 2005.

[31] LaValle SM. *Planning algorithms:* Cambridge University Press; 2006.

[32] Jarris RA. Collission-free trajectory planning using distance transforms. *Mechanical Engineering Transactions of the IE Australia*. 1985;10:187.

[33] Latombe J-C. *Robot motion planning*: Springer Science & Business Media; 2012.

[34] Khatib O. Real-time obstacle avoidance for manipulators and mobile robots. *Autonomous robot vehicles*: Springer; 1986. p. 396-404.

[35] Trevisan M, Idiart MAP, Prestes E, Engel PM. Exploratory navigation based on dynamical boundary value problems. *Journal of Intelligent and Robotic Systems*. 2006;45:101-14.

[36] Masoud SA, Masoud AA. Motion planning in the presence of directional and regional avoidance constraints using nonlinear, anisotropic, harmonic potential fields: a physical metaphor. *IEEE Transactions on Systems, Man, and Cybernetics-Part A: Systems and Humans*. 2002;32:705-23.

[37] Bräunl T. *Embedded robotics: mobile robot design and applications with embedded systems*: Springer Science & Business Media; 2008.

[38] Kavraki L, Svestka P, Overmars MH. *Probabilistic roadmaps for path planning in high-dimensional configuration spaces*: Unknown Publisher; 1994.

[39] Geraerts R, Overmars MH. A comparative study of probabilistic roadmap planners. *Algorithmic Foundations of Robotics V*: Springer; 2004. p. 43-57.

[40] Schwartz JT, Sharir M. *Algorithmic motion planning in robotics*. Algorithms and Complexity: Elsevier; 1990. p. 391-430.

[41] Schwartz JT, Sharir M. On the "piano movers'" problem I. The case of a two-dimensional rigid polygonal body moving amidst polygonal barriers. *Communications on pure and applied mathematics*. 1983; 36:345-98.

[42] Perez-Lozano T. Spatial planning: A configuration space approach. *IEEE Trans Computers*. 1983;32.

[43] Huang Y, Cao Z, Hall E. Region filling operations for mobile robot using computer graphics. *IEEE*. p. 1607-14.

[44] Ollis M, Stentz A. First results in vision-based crop line tracking. *IEEE*. p. 951-6.

[45] Land S, Choset H. *Coverage path planning for landmine location*.

[46] Laubach SL, Burdick JW. An autonomous sensor-based path-planner for planetary microrovers. *IEEE*. p. 347-54.

[47] Langer RA, Coelho LS, Oliveira GHC. K-Bug, a new bug approach for mobile robot's path planning. *IEEE*. p. 403-8.

[48] Sciavicco L, Siciliano B. *Modelling and control of robot manipulators*: Springer Science & Business Media; 2012.

[49] Sankaranarayanan A, Vidyasagar M. A new path planning algorithm for moving a point object amidst unknown obstacles in a plane. *IEEE*. p. 1930-6.

[50] Kamon I, Rivlin E. Sensory-based motion planning with global proofs. *IEEE transactions on robotics and automation*. 1997;13:814-22.

[51] Noborio H, Nogami R, Hirao S. A new sensor-based path-planning algorithm whose path length is shorter on the average. *IEEE*. p. 2832-9.

[52] Magid E, Rivlin E. CAUTIOUSBUG: A competitive algorithm for sensory-based robot navigation. *IEEE*. p. 2757-62.

[53] Kamon I, Rimon E, Rivlin E. Range-sensor based navigation in three dimensions. *IEEE*. p. 163-9.

[54] Kriechbaum KL. *Tools and algorithms for mobile robot navigation with uncertain localization*. 2006.

[55] Stentz A. *Optimal and efficient path planning for partially known environments*. Intelligent Unmanned Ground Vehicles: Springer; 1997. p. 203-20.

[56] Horiuchi Y, Noborio H. Evaluation of path length made in sensor-based path-planning with the alternative following. *IEEE*. p. 1728-35.

[57] Lozano-Pérez T, Wesley MA. An algorithm for planning collision-free paths among polyhedral obstacles. *Communications of the ACM*. 1979;22:560-70.

[58] Spandl H. *Das HALMOR System. Maschinelles Lernen* [*Das HALMOR system. Machine learning*]: Springer; 1992. p. 63-97.

[59] Lengyel J, Reichert M, Donald BR, *Greenberg DP. Real-time robot motion planning using rasterizing computer graphics hardware*: ACM; 1990.

[60] Lozano-Perez T. A simple motion-planning algorithm for general robot manipulators. *IEEE Journal on Robotics and Automation*. 1987;3:224-38.

[61] Udupa SM. *Collision detection and avoidance in computer controlled manipulators*. 1977.

[62] Donald BR. *A search algorithm for motion planning with six degrees of freedom*. Artificial Intelligence. 1987;31:295-353.

[63] Dorst L, Trovato K. Optimal path planning by cost wave propagation in metric configuration space. *International Society for Optics and Photonics*. p. 186-97.

[64] Khatib O, Le Maitre JF. *Dynamic controlof manipulators operating in a complex environment*.

[65] Barraquand J, Latombe J-C. Robot motion planning: A distributed representation approach. *The International Journal of Robotics Research*. 1991;10:628-49.

[66] Koditschek D. Exact robot navigation by means of potential functions: Some topological considerations. *IEEE*. p. 1-6.

[67] Pavlidis T. Contour filling in raster graphics: *ACM*; 1981.

[68] Yufka A, Parlaktuna O. *Performance comparison of the BUG's algorithms for mobile robots*. p. 416-21.

[69] Brandão AS, Sarcinelli-Filho M, Carelli R. An analytical approach to avoid obstacles in mobile robot navigation. *International Journal of Advanced Robotic Systems*. 2013;10:278.

[70] Secchi H, Carelli R, Mut V. *Discrete stable control of mobile robots with obstacles avoidance*. p. 405-11.

[71] Vidyasagar M. *Nonlinear systems analysis*: Siam; 2002.

[72] Fakoor M, Kosari A, Jafarzadeh M. Revision on fuzzy artificial potential field for humanoid robot path planning in unknown environment. *International Journal of Advanced Mechatronic Systems*. 2015;6:174-83.

[73] Mobayen S, Javadi S. Disturbance observer and finite-time tracker design of disturbed third-order nonholonomic systems using terminal sliding mode. *Journal of Vibration and Control*. 2017;23:181-9.

[74] Bellman R. A Markovian decision process. *Journal of mathematics and mechanics*. 1957:679-84.

[75] Kolobov A. Planning with Markov decision processes: An AI perspective. *Synthesis Lectures on Artificial Intelligence and Machine Learning*. 2012;6:1-210.

[76] Fakoor M, Kosari A, Jafarzadeh M. Humanoid robot path planning with fuzzy Markov decision processes. *Journal of applied research and technology*. 2016;14:300-10.

[77] Sheskin TJ. *Markov chains and decision processes for engineers and managers*: CRC Press; 2016.

[78] Puterman ML. *Markov Decision Processes: Discrete Stochastic Dynamic Programming*: John Wiley & Sons; 2014.

[79] Hu Q, Yue W. *Markov decision processes with their applications*: Springer Science & Business Media; 2007.

ABOUT THE AUTHORS

Hadi Jahanshahi

Researcher, Department of Aerospace Engineering,
University of Tehran, Iran
Email: hadi_jahanshahi@ht.ac.ir

Hadi Jahanshahi is a holder of BSc in mechanical engineering and MSc in aerospace engineering from the University of Tehran. Hadi has published over 20 peer-reviewed papers in reputed international journals, besides 3 text-books in optimization and robotics. Hadi has selected as the best student in the bachelor's program and the best student and researcher in the master's program. Additionally, his master's thesis is selected as the best thesis in space engineering by Iranian Aerospace Society. Moreover, he has awarded a fully funded PhD in mechanical engineering from K.N Toosi University of Technology. His main research interests include optimal and nonlinear control, numerical computational methods, robotics, and complex systems.

Naeimeh Najafizadeh Sari

PhD Student, Department of Mechanical Engineering,
University of Manitoba, Canada
Email: naeimeh.najafi@gmail.com

Naeimeh Najafizadeh Sari was born in Tehran, Iran. She received her BSc degree in Aerospace engineering from the Amirkabir University of Technology (Polytechnic of Tehran) in 2015, and the MSc in Aerospace engineering from the University of Tehran, in 2018. She is currently a PhD student in mechanical engineering at the University of Manitoba. Her current research interests include control systems, robotics, space systems, system engineering and product assurance, and hybrid vehicles.

Viet-Thanh Pham

Assistant Professor, Faculty of Electrical and Electronic Engineering,
Phenikaa Institute for Advanced Study (PIAS),
Phenikaa University, Vietnam
Email: thanh.phanviet@phenikaa-uni.edu.vn

Viet-Thanh Pham received the degree in electronics and telecommunications from the Hanoi University of Technology, Vietnam, in 2005, and the Ph.D. degree in electronics, automation and control of complex systems engineering from the University of Catania, Italy, in 2013. His scientific interest includes nonlinear control, applications of nonlinear systems, mobile robots, analysis and design of analog circuits, and FPGA-based digital circuits.

Roya Khajepour

Researcher, Center of Excellen in Robotics and Control,
Advanced Robitics and Automates System (ARAS) Lab,
Department of Mechanical Engineering,
K.N. Toosi University of Technology, Iran
Email: r.khajepoor@gmail.com

Roya Khajepour received the B.Sc. degree in Mechanical Engineering from Amirkabir University of Technology, Iran, in 2011. Then she attended the Faculty of K. N. Toosi University of Technology in order to continue her studies as a M.Sc. student of Mechatronics Engineering. Meanwhile, she pursued her research at Center of Excellent in Robotics and Control; ARAS Laboratory. Recently, she has graduated from K. N. Toosi University of Technology. She has an extensive set of research interests, including applications in robotics, system identification, control, machine learning, and mobile robots. She has been research assistant in Advanced Robotics and Automated System Laboratory (ARAS LAB) for two years.

Christos Volos

Assistant Professor, Department of Physics,
Aristotle University of Thessaliniki, Greece
Email: volos@physics.euth.gr, christov@auth.gr

Christos Volos received the Physics Diploma degree, the M.Sc. degree in Electronics, and the Ph.D. degree in Chaotic Electronics from the Physics Department, Aristotle University of Thessaloniki, in 1999, 2002, and 2008, respectively. He is currently an Assistant Professor with the Physics Department, Aristotle University of Thessaloniki, Greece and a member of the Laboratory of Nonlinear Circuits – Systems & Complexity (LaNSCom). His current research interests include, chaos, nonlinear systems, memristor and memristive systems, design of analog and mixed signal electronic circuits, neural networks, chaotic electronics and their applications (secure communications, cryptography, robotics), chaotic synchronization and control. Dr. Christos Volos has more than 200 publications as author or co-author in international journals, book chapters, and international conference papers. He has edited 7 books related with nonlinear circuits and systems. Furthermore, he is member of 8 international journal's editorial boards and he has been guest editor in 15 special issues of international journals. Also, he is closely associated with several international journals as a reviewer. Finally, Dr. Christos Volos has participated in several research programs related with chaos and its applications.

INDEX

C

D

E

F

G

H

I

K

L

M

N

O

P

Q

R

U

V

W

Z